George Washington Carver

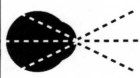
This Large Print Book carries the
Seal of Approval of N.A.V.H.

George Washington Carver

Inventor and Naturalist

Sam Wellman

Thorndike Press • Thorndike, Maine

©1998 by Sam Wellman

All rights reserved.

Published in 2001 by arrangement with Promise Press, an imprint of Barbour Publishing, Inc.

Thorndike Press Large Print Christian Fiction Series.

The tree indicium is a trademark of Thorndike Press.

The text of this Large Print edition is unabridged.
Other aspects of the book may vary from the original edition.

Set in 16 pt. Plantin by Warren S. Doersam.

Printed in the United States on permanent paper.

Library of Congress Cataloging-in-Publication Data

Wellman, Sam.
    George Washington Carver : inventor and naturalist / Sam Wellman.
        p. cm.
    Originally published : Urichsville, Ohio : Barbour Pub., c1998.
    Includes bibliographical references.
    ISBN 0-7862-3065-7 (lg. print : hc : alk. paper)
    1. Carver, George Washington, 1864?–1943. 2. Afro-American agriculturists — Biography. 3. Agriculturists — United States — Biography. I. Title.
S417.C3 W45 2001
630′.92—dc21
    [B]   00-048903

George Washington Carver

one

The sun had peeked up over the dark Ozark humps to the east, warming George on the front porch of the main house on the Carver farm — not that George needed much warming that morning in 1871. Churning butter for what seemed forever did a real good job of that already.

"Come on, George!" snapped Jim impatiently, his cane fishing pole over his shoulder. "I fed all the livestock."

"I have to finish this butter," explained George needlessly to his brother, and he churned the dasher even harder to make his point.

"The fish won't be biting if it gets too warm. Pour some hot water in the dasher."

"Oh, no . . . ," whispered George.

He had done that once. Hot water set the butter up fast all right, but the butter was telltale white and foamy! Jim would just have to wait a little longer. Aunt Sue was strict enough, with a hickory switch applied one hard time for disobedience, but Uncle Mose was downright rough.

Even if Uncle Mose didn't whop George a couple of times with a leather belt, he could ice George's very soul with those hawkish, disapproving eyes.

"I think it's set up," said George finally.

"When you take it inside to Aunt Sue, make sure you get some of those fat trimmings for bait," reminded Jim.

Soon the boys were trudging barefoot down to the creek. George's brother Jim was a strapping boy for twelve years of age. He wore shirts and pants of wool and leather in the winter or of cotton and linen in the summer, just like Uncle Mose did. George was still what folks called a "shirt-tailed boy," wearing a wool or cotton shirt that came down to his knees. But maybe Jim was older than twelve, reflected George. George didn't know for sure his own age, either. He might have been eight. When he asked Aunt Sue she began to cry because when George was a baby he and his mother Mary and his sister Melissa had been grabbed off the farm by some bushwhackers. When George asked Uncle Mose he just shook his head and told how trying to get Mary back from the bushwhackers cost him his best horse, Pacer. All he got for Pacer was George. The bushwhackers had decided they didn't want a sick, bawling baby. It happened back in what folks called the Civil War, although how a war could be civil — in the sense defined in George's beloved *Webster's Spelling Book* — he couldn't fathom. And he tried to fathom everything.

"Ever wonder if you could mix flowers together, Jim?" he asked, after baiting his line with a cricket.

"Mix flowers?" mumbled Jim.

"Remember how Uncle Mose mixed that tan and white hound with a black hound and he got tan and white and black pups?"

"I guess I do."

"Well, why can't you do that with flowers?"

Jim looked at George numbly. "I guess you'd have to know how to mix flowers."

George had thought about mixing flowers a lot. One morning after Sunday school at the Locust Grove schoolhouse a mile away, he had stayed to hear the church service. One Bible reading struck him like a thunderbolt:

And if some of the branches be broken off, and thou, being a wild olive tree, wert grafted in among them, and with them

9

partakest of the root and fatness of the olive tree . . .

George had been overwhelmed. Graft a wild olive tree into an orchard tree? Was that what the Bible reading meant? He had raced back to the farmhouse to find Uncle Mose. "Sure," Uncle Mose had replied, "folks do graft branches of fancy varieties of fruit trees into already grown trees. Been doing it since Bible times probably. But no," Uncle Mose had continued, "that was not like getting black and white and tan puppies or mixing flowers. . . ."

"Think they teach how to mix flowers in school?" George asked Jim, who flipped his line into a dark recess of the creek.

"Sunday school?" Jim concentrated. "Seems like they talk about Jesus a lot. . . ."

"No, I don't mean Sunday school. I mean regular everyday school."

"It doesn't matter, because you can't go to regular school."

"Uncle Mose said it was all right with him."

"But the regular school won't take colored* folks like us. Why don't you just keep

---

* The term "colored is historically accurate and

10

studying that old speller Aunt Sue gave you?"

"Why won't the school take folks like you and me?"

Jim couldn't answer that, and he grumbled, "All your blabbering is scaring the fish away." He rebaited his hook with a worm he dug up near the creek. But after no success with that he said, "Why don't you go down to the regular school and sit outside the window and listen? Or have you done that already?"

"I can't see what they're reading."

"Well, why would you want to go to school anyway?"

"I want to learn how to mix flowers. I want to know the names of things."

"We've got everything right here on the farm," blurted Jim in exasperation. "Uncle Mose says we don't buy anything but coffee and tobacco. Everything else we make right here on the farm: flax and hemp

---

used without negative connotations. Its use maintains consistency with Mr. Carver's quotes and those of his contemporaries. The blatantly derogatory terms "nigger" and "darkey," used sparingly elsewhere in this book, are also historically accurate, but are retained only as a demonstration of the hostility and discrimination faced by Mr. Carver during his lifetime.

for clothes and cord. Cattle for beef and milk and leather to make harnesses, shoes, and belts. Pigs, ducks, chickens, and eggs for eating. Corn for the stock and feeding ourselves. Trees that drop apples, peaches, walnuts, hazelnuts, and pecans right into your outstretched hand. Why, we even have honey from dozens of bee hives back in the timber. And speaking of timber, is there better wood for making furniture than the oak and walnut we have here? And Uncle Mose raises fine horses and hounds to sell. Wake up, George. Everything you would ever need is right here on the farm."

"Seems like something's missing," mumbled George, and resisted the impulse to tell Jim just how thirsty he was for knowledge. Somehow he had to quench his burning desire for knowledge.

"Well, it's May and school will soon be out for everybody anyway," snapped Jim.

Maybe if George had worked as a hand on the farm like Jim he would have felt the same way Jim did. But George was a houseboy. He wasn't built husky like Jim. George seemed as thin as a pipe stem and sick a lot. "Boy won't live to see twenty-one," whispered some folks, though Jim often muttered "Aw, he just *looks* sick." Sometimes George felt like Aunt Sue

12

thought of him as her adopted daughter! He dusted the furniture, not that there was much besides a table, benches, a bed, and two spinning wheels — one for fine linen and one for cotton. He changed the straw up in the loft where he and Jim slept. He swept and scrubbed the split-log floors. He peeled vegetables. He ironed tablecloths. He pampered the fireplace. He tended the Dutch oven that sat over the fire. He washed the pewter, dishware, and utensils, then arranged them on the table. He could knit with four needles. He even crocheted. And he was very good at all those things. Oh, he went outside, too. But only to hang wet clothes. To chop firewood. To churn butter. To garden next to the house.

"Gardening is a blessed thing," he admitted to himself.

Things about plants just stuck in his mind like glue. In the month of May like it was now he knew, almost without thinking, that he had to weed the cabbage, turnips, cauliflower, carrots, potatoes, and the fat onions. And the ground had to be prepared for tomatoes, squash, pumpkins, beans, cucumbers, green peppers, and, best of all, several kinds of sweet melons. And to a diligent gardener, the bounty was already being harvested in May: lettuce, radishes, green onions, and spinach. Of course he wasn't allowed to go out in the fields to plant the sweet corn. Jim did that, under Uncle Mose's stern eyes.

"Someday I'll go out in the fields and show them just how to raise corn and flax and hemp," he declared as he plucked up green onions.

Often Aunt Sue would come outside and sigh, "The chores are pretty much in hand, George. Why not go see your own plants?" And George would dash off into the timber. His head swam from the sight of the riches. The trees were most obvious and George felt privileged to know their names. There were two kinds of hickory: bitternut and shagbark. Only the shagbark nuts tasted good. Of the several kinds of oak, the white and red kinds were good for making furniture. The blackjack and pin oak were only good for fence posts and such. But oaks were good for tanning leather and dying cloth, too. And who could ignore all the good things about walnut and black cherry trees? And wild grapes, plums, chokecherries, currants, raspberries, blackberries, and sugarberries could be found in clearings and on slopes below the timber. There was such wealth there!

14

"God, thank You for giving all this to us," he said, remembering his Sunday school lessons.

George had been talking to God ever since a neighbor boy showed him how to pray. After the boy left, George went into the barn where he could be alone and prayed with all the love he could muster to God, the Creator. Something happened to George at that moment. If he had been a bird he would have soared to the heights. Instead he seemed to become rooted to all of God's creations. He dug in the leaf litter and the rich soil to find dozens of kinds of nuts and mushrooms. Grubs and spiders and centipedes didn't bother George at all. All around — on and under the leaf litter and rotting logs, in the brush, up in the trees — critters swarmed. Beetles and butterflies. Toads, tree frogs. Dashing lizards and clumsy box turtles. Woodchucks, voles, and mice. Raccoons, possums, rabbits, and squirrels. Hooty owls. Whippoorwills, woodpeckers, wrens, warblers. A red-shouldered hawk perched above and watched every movement as intently as George. Some seemed only to visit the forest like George. Soaring high overhead was some kind of hawk larger than the one sitting in the timber. And some birds were only passing by. A pintail, a duck with one long tail feather, scooting north had been a sure sign of spring.

In some places great tan rocks swelled out of the hillside. The rocks were so hard George could scarcely scratch them with the small trowel he dug with. The rocks had the strangest things on them. There were forms that looked like stalked plants and fans and snails and such. Living things. But they were rock themselves! They were wonderful puzzles, but George didn't linger too long around such wonders. Rocks were a favorite place for snakes. One close call from a rattlesnake stuck in George's memory. These rocky places also had caves — and caves had bears. George knew that for certain because Uncle Mose had shot a big black bear once. Bears could run fast and they could climb trees, so George was no fool around the rocks. Still, the puzzles in the rocks drew him like a magnet drew a hobnail.

Often George would pick up a rock or plant or a bug to take back to the farm. There he quizzed folks about them. About as often as not Uncle Mose and Aunt Sue would not know a name for his curiosity. So George made up his own name. He just

16

felt lucky if Aunt Sue didn't scream, "Get that thing out of the house, George!" Besides naming his treasures he kept them all together in what Uncle Mose called his "collection." George was very proud of his collection.

"But that is not all," he told himself with delight so great he felt he would burst.

In the timber he had a secret garden, small but oh so special. He had already figured out how to trick the seasons. If he covered shade-loving plants in the winter they just kept right on growing. Heat from the earth just didn't go wafting off into the air but gathered under his cover to bathe his tender plants in warmth. Right in the dead of winter he could look at delicate ferns, wonderfully shaped gourds and, most amazing of all, several kinds of blossoms. At first his flowering plants looked similar, yet his intense interest revealed they were not the same at all. The leaves differed. Things jutting out of the center of the flowers differed in shape and color. And the petals testified to that fact, too. The number and shape of the petals varied, as did their color. One kind was yellow. Another was blue, another purple. He plucked a flower and took it to Aunt Sue. "Is it a violet?" she mused. She wasn't

17

sure. Who could tell George about such things? Was it possible no one in the whole wide world knew? Puzzlement did not spoil his pleasure in viewing his flowered treasures as icicles hung overhead from the trees.

His triumphs were not confined to his secret garden in the timber. One spring Uncle Mose, half-joking, asked him, "What's wrong with that apple tree, George?" As nimble as a squirrel George climbed deep into the foliage of the sick tree. Soon he yelled, "I found a branch all clogged with worms, Uncle Mose. If I had a saw I'd cut it right off." After George sawed the branch off and it tumbled to the ground Uncle Mose said, "Shot right through with caterpillars of codling moths." After the apple tree recovered, George's reputation blossomed, too. Even though the kids started teasing him about being the "plant doctor," he was often summoned to a neighbor's farm. Of course, although he wanted to, he knew no "shirt-tailed boy" could advise the touchy farmers on how to grow corn or flax, but he could talk to their wives about their flowers.

That same spring over at the Baynhams', he was asked to diagnose Mrs. Baynham's sick roses. He looked them over no more than a few seconds and asked for a spade. Then he carefully dug deep around them, and protecting the root balls he replanted them in full sun on the south side of her large brick house. "Roses surely do love sun, ma'am," he explained. And he loved roses, he told himself. She gave him a three-cent piece and the honor of standing in her parlor. George was shocked by paintings displayed there. He had never seen a painting before. There were no paintings at all on the Carver farm. Was it possible for a person to paint faces and scenes of trees and rivers that looked so real? Obviously. After that, George thought a lot about drawing and painting, too. Should he try it himself? There wasn't much he couldn't do if he set his mind to it. He already knew something about yellow and brown dyes from oaks. There was a pokeberry growing wild that Uncle Mose said never to eat. Now George knew it wasn't worthless. The juice of the berries stained things blue or red. Maybe the juice could be used for painting. Did every plant have some hidden use?

But at supper one night he had another surprise. "George is going to need some pants," commented Uncle Mose.

19

"Pants?" blinked Aunt Sue.

"Time he went into Neosho," grunted Uncle Mose.

"Neosho!" George dropped his fork. "Me?"

two

Neosho wasn't like nearby Diamond Grove, which wasn't much more than a livery and a general store so small two or three folks seemed a crowd. Neosho was a town of thousands of folks about ten or so hilly miles south of Diamond Grove. Brother Jim had been walking there alone for the past two years. Now George was deemed old enough to go there. One spring dawn he stood on the front porch dressed in a clean shirt and his new cotton pants. His boots were slung over his shoulders by their tied rawhide laces. He would put them on as he approached Neosho. He sure couldn't wear out his nice boots just hoofing it through the hill country.

"Why aren't you carrying your boots?" George asked Jim, who was grinning at him.

"He's not going," answered Uncle Mose.

"You mean I'm going alone?" blurted George.

"Well now, doesn't Jim go into Neosho alone?" asked Uncle Mose calmly.

21

"I just figured . . ."

"Two scalawags do ten times the mischief of one scalawag!" explained Uncle Mose gruffly.

"Yes, sir," mumbled George, now thoroughly unsettled.

"Don't forget to take some of your savings if you want," reminded Uncle Mose. "Buy something useful."

"I've got eight cents with me, sir." George reached in his new pockets — what a boon pockets were! — and plucked out two coppery pennies and two gray three-cent pieces. In his box in the loft he had two half-dimes, worth five cents each but so pretty in their silvery tininess he couldn't part with them. He had a silver dime, too. He sure wasn't going to part with that treasure.

"Here's a silver dollar," grunted Uncle Mose. "Put it in your other pocket. We need a can of coffee. Make sure it's 'Rio' coffee. Don't let them stick you with that 'Java' coffee."

"And don't forget your grub," reminded Aunt Sue.

"I've got my corn dodgers, ma'am."

George held up a pouch and withdrew it in a flash. He didn't want Aunt Sue to know he had meticulously sliced open each crispy cornmeal ball and inserted a nice wedge of sidepork. He hadn't asked her permission. Thinking about his deception made him so edgy he hardly remembered their farewells and warnings as he trudged off toward Diamond Grove in the cool dawn. Besides that, his mind was on eating his delicious invention on the trail.

"A man needs his grub," he said after only a few minutes of walking and he ate one of his mouth-watering dodgers.

It was rule of thumb that a man could walk about four miles in one hour. So the journey to Neosho should have taken between two and three hours. But this was George's first time alone on the trail and there was just too much to see. He hadn't walked beyond Diamond Grove more than a mile before he saw a trumpet creeper vining on the bank and bearing flowers. Its orangey conical blossoms shouldn't have appeared for another month or so. Naturally George had to inspect it very carefully to find out just why it bloomed so early. Was it the soil? Was it the amount of sun it received? Did water pool up hereabouts? Finally he decided it must be the extra sun it got on a south slope.

His deliberations along the trail slowed him down. He was walking only about two miles every hour. Besides that, he wasn't so sure how much he really wanted to be in Neosho. Someone, maybe it was Uncle Mose's ornery nephew Dick, had said Neosho had been a stronghold for a while of the Confederate troops. Some of these white folks weren't nice like Uncle Mose and Aunt Sue. So, if George understood it right, that meant some white folks in Neosho wanted to keep colored folks as slaves. Would some giant bushwhacker — even meaner than Dick — grab him? The sudden thought of the villains who stole George's mother Mary and sister Melissa stung him like a hornet. Still, George's brother Jim went to Neosho and he never complained.

"Sometimes I feel like Jim," chirped George. He often talked aloud among God's plants and animals. "The farm has everything anyone would ever want in life, so why leave? But at other times I feel like there's something missing. Will I find it in Neosho? Whoa!" He eyed tracks impressed into the dust on the trail. Each paw print had four toes, with no claw marks. "Bobcat, for sure . . ."

It was noon by the time George descended into the valley of Shoal Creek. After he crossed Shoal Creek he put his boots on and ascended into Neosho. He felt in his pocket for the silver dollar for the hundredth time. If he lost it he might spend the rest of his life paying back Uncle Mose. The first sight he saw on his ascent was a nursery with thousands of small saplings all bundled with their root balls bagged. A sign advertised prices of the planting stocks: apple trees, $5 a hundred; pear trees, $25 a hundred; grape vines, $5 a hundred; cherry trees, $20 a hundred; and flowering shrubs, $1.50 a dozen.

George gulped. "There's a lot of money in raising plants in a nursery." He saw a glass-paned house there too and understood its use immediately. The glass multiplied the power of the sun to warm its plants. "It's called a 'greenhouse,'" he noted, reading a sign hung over its door.

Neosho's size overwhelmed him. On the outskirts some houses were split-log houses like Uncle Mose's two houses, but toward the center of town the houses were white frame, even brick like the Baynhams'. And not just one or two houses. Too many to count. And belfries signaled churches. Streets ran this way and that way. There were buckboards and buggies and carriages everywhere. Some of the carriages were so fancy with nice leather tops

25

and leather seats, George couldn't believe his eyes. Mountings of lamps and such were mostly brass but some looked to be ivory, even silver. And the colors of the carriages! Black, plum, blue, yellow, and a dozen kinds of green, some as pale and gray as olive leaves, some as deep green as an oak leaf. Some of the greens shaded into blue, some into yellow. How he admired such mixing of colors. Many of the colors were beyond George's power of description. And that realization always stung him like a hornet, too.

Still, he was delighted. "So this is what awaits a fellow away from the farm!"

He noticed dust billowing above the buildings. Then he saw a stagecoach thundering across an intersection. Six horses paired in three "spans" pulled it. The driver wielded a whip that must have been twenty feet long. It cracked like lightning. George knew something about horses because Uncle Mose raised horses. Good ones, too. Some of Uncle Mose's horses sold for more than one hundred dollars, folks said. But in Neosho there was horse-flesh George had heard folks talk about but had never seen before. There were ponies. There were massive dobbins. Broomtails. Cobs. And what a variety of colors: dun, sorrel, albino, skewbald, Appaloosa, roan, bay, pinto, buckskin, piebald, buttermilk, smokey, claybank, and chestnut!

"Who do you belong to, boy?" asked a white man who didn't appear very friendly.

"Uncle Mose Carver, sir," answered George, falsely, but somehow knowing that was the right way to answer this scowling man.

"Good thing for you I've heard of Moses Carver. Well, what you looking for, boy?"

"General store, sir."

"Right down that street over there," he said, pointing. "And you keep your eyes off my horse. You better learn to keep your black eyes down in the dirt where they belong, boy."

George's heart was hammering. Now he was afraid to go into the general store. Still, he knew everyone back on the farm would ask what he saw. He couldn't just turn around and run. Jim had said, with a peculiar pained look, that a few folks in Neosho might be a bit unfriendly. The street with the general store was wide and lined with tall, flat-fronted buildings. George stepped up on the boarded sidewalk and read all the marquees. There were hardware and clothing stores, law offices, a land office, a surveyor's office, saloons, a tobacco shop, a bakery, a boot maker's shop, a hotel, a cafe, a drug store, a paint store, and what all else George couldn't take in all at once.

He then watched folks on the street for a while, slyly now that he had been warned. A good trick was to appear to be looking at goods displayed inside the general store window but actually be looking at the reflections from the street. George was amazed at the fancy way folks dressed. Men in spanking-new felt hats, with bright ribbons around their necks that tucked down into their vests. Women in great billowing dresses and wide, loppy hats. And there were so many materials in the dresses. George recognized calicos and poplins easy enough. But what were these others fabrics? Were they the muslin and taffeta and watered silk and velvet and satin he had heard about? Their colors were rich reds, blues, browns, and plums. Once again he saw a variety of greens that defied his powers of description. And who could not marvel at how the dresses were festooned with lace, sashes, nets, ribbons, billows, and ruffles?

"What do you want, boy?" This was

from a white man who had just stepped out the front door of the general store. He wore an apron.

"I don't know for sure, sir."

"Got any money on you?"

"Eight cents, sir."

"Come on in then. I'm a clerk in here."

George wasn't in the general store two seconds before the clerk told him not to touch anything. So George just looked. The clerk noticed his confusion and rattled off things George could buy for eight cents. There were many things he could afford. There were several kinds of candy sticks in jars. There were twangy mouth harps. There were bobbers and hooks for a fishing line. Rubber balls. Glass marbles. Finally George saw something he really wanted.

"How much is that, sir?"

The clerk blinked. "Why, do you want to buy a crochet hook for your dear old mammy, boy?"

"I'd like to buy it, sir," answered George evasively.

Emboldened by his purchase he checked out the prices of other merchandise in the store; if the merchandise wasn't marked he even asked the clerk now. A pair of copper-toed shoes for a boy cost a dollar; a man's good boots, four dollars. A lot of the food was in metal containers the clerk called "cans." Corn or beans or peaches in a can cost twenty-five cents, but George couldn't imagine why anyone would buy them in a can. Regular corn was fifty cents a bushel. Potatoes were eighty cents a bushel. Eggs were twenty cents a dozen. Fresh ham was eighteen cents a pound. Syrup was a dollar a gallon. The best white sugar was a dollar for seven pounds. George filed all these prices in his memory.

"Uncle Mose never talks about the price of anything but horses," muttered George.

"What'd you say, boy?" asked the clerk.

"I need a can of coffee, sir. The Rio kind."

The clerk laughed. "It's one dollar a can, boy. Did you ever see a dollar?"

George politely slid his silver dollar onto the counter. "That will be fine, sir. The coffee is for Mister Mose Carver over yonder by Diamond Grove."

Finally, George left the general store with his crochet hook and a can of Rio coffee. There seemed little else that could top what he had already seen so he started back to the farm. But near the outskirts of Neosho he was amazed to see colored children flooding from a split-log cabin. Ten. Twenty. Thirty. Forty. Lord Almighty, too many to count! They were yammering and squawking like a gaggle of geese. They were happy, too, oh so happy. They stampeded past George, ignoring his gaping jaw and his questions. Then George saw that a colored man stood in the doorway.

"Is this here a school for colored folks?" called George, his heart pounding wildly.

"It surely is for colored children," called the man, "and why aren't you in school, youngster?"

"I live over yonder near Diamond Grove," answered George weakly.

George didn't know whether to fly or cry all the way back to the farm. Of all his marvelous discoveries this day the school was best of all. But the opportunity jolted him, too. Could he leave the farm to come to Neosho to go to school? Oh, he knew Uncle Mose would let him leave. Uncle Mose had told him that a long time ago. George was as free as a bird. He could stay as long as he was willing to work there, or he could leave. But could he leave? He loved the farm. He loved the timber. He loved his secret garden. He loved Uncle Mose and Aunt Sue. Uncle Mose treated him like a son, he supposed. He whipped

31

George, advised him, worried over him. So did Aunt Sue. And how could George leave his brother Jim? Could he even imagine fishing without Jim? Or not sleeping with Jim by his side at night? And what about all the good times they had with the neighbors and all the other Carvers? What about the times they gathered to husk corn, pare apples, throw up a building? Then afterward they all sang and danced. Folks said Uncle Mose was about the best fiddler in Newton County. And how George loved to imitate Uncle Mose by stomping his foot and fiddling on a cornstalk with a horsehair bow! And no one ever said, "You better keep your black eyes down in the dirt where they belong, boy." No, sir. Not if any of the many Carvers were around.

"Uncle Mose would horsewhip that scoundrel in a hurry!" spouted George and startled a lizard off the trail.

And Aunt Sue was so kind to him. Could George live without her hugs? The more he thought about leaving the farm the more upset he became. Where would he stay in Neosho? How would he eat? He didn't know a soul there. But didn't the bird have to leave the nest? How would he ever learn everything he wanted to learn if he didn't go to school? Because just raising plants wasn't enough. Maybe even having a greenhouse like the one he saw in Neosho wouldn't be enough. He wanted to know all the names of everything about every plant and how each part worked and when and why. And what about his newly discovered painting? That couldn't be done with the resources of a farm. He concluded that after trying to mix concoctions of pokeberry juice and such. For some reason he remembered a parable from Sunday school. He rarely forgot anything he heard. It kind of puzzled him when folks said, "By gum, I can't remember." But sometimes just remembering wasn't enough. A fellow had to twist and turn what he remembered, to find deep meanings.

In the parable a master had three servants. He gave each one some money, as much as their ability merited. To one he gave five "talents." To another two. To the least able servant, one talent. Later the master learned the most able had doubled his five talents. "Well done, thou good and faithful servant," the master said. The servant was given many new responsibilities and had the joy of the Lord. The servant to whom the master had given two talents

33

also doubled his money. This servant too was rewarded. But the servant least able, the one who had been given one talent, was afraid to lose it and buried his money in the ground. The rest of the parable chilled George because the master said to the least able servant:

Thou wicked and slothful servant, thou knewest that I reap where I sowed not, and gather where I have not strawed. . . . Take therefore the talent from him, and give it unto him which hath ten talents. For unto every one that hath shall be given, and he shall have abundance: but from him that hath not shall be taken away even that which he hath. And cast ye the unprofitable servant into outer darkness: there shall be weeping and gnashing of teeth.

The parable seemed very harsh to George. But he understood it plain enough. God did not want folks to neglect the abilities He gave them. Perhaps God had led George past that log schoolhouse just precisely when the children were coming out, screaming and hollering at the top of their lungs, so George would know it was there. Yes, God's purpose seemed more certain every second George thought about it. God had given him a talent, too. Now what was he going to do with it? Go back and bury it on the farm? Because the farm had already been given to George. Uncle Mose had made the farm. He had come from Ohio, married Susan Blue in Illinois and settled right here in Missouri back in '38. This farm was the ten talents of Uncle Mose. For sure, if George stayed on the farm God would take his one talent and give it to Uncle Mose. For George, was this farm "outer darkness"? Would there be "weeping and gnashing of teeth"? Maybe George wasn't twisting and turning the facts just right yet, but he had the general idea, he was sure of that. George must leave the farm to go to school. And yet leaving the farm to go to school in Neosho worried him. How would he eat? Where would he sleep?

Once he said brightly, "I'll try school in Neosho for just a spell, just as soon as the field corn is in."

But he didn't go. Years of comfort on the Carver farm passed. Yet George was troubled. By 1875 his deepest longings gnawed at him, still unfulfilled. More and more he prayed for God's guidance. Then one night something wonderful happened. . . .

three

George had a "vision."

He had been thinking a lot about God and just how personal God got with a fellow. And he had been wondering just how much a fellow should trust God. Because to just up and go off to Neosho alone a fellow surely needed to trust God. Then one night up in the loft George dreamed of a nice pocketknife lying out in the field at the base of three cornstalks, near the rind of a half-eaten melon. There wasn't much he wanted more than a nice pocketknife. After George woke up the next morning he raced to the cornfield where Uncle Mose tried to hide his melon patch. Sure enough, near three corn stalks was a carved-up melon.

"And a pocketknife!" he gasped.

The entire episode was choked with symbols. At the foot of three cornstalks. At the foot of the three crosses of Calvary. And the melon. Was that Christ's sacrifice? And the knife. A sign. Maybe George couldn't interpret it all, but surely it was a message from God: "I am with you." George trudged back to the house, with not just a pocketknife, but a stronger resolve to go to Neosho.

"I'm going off to school in Neosho," he said to Uncle Mose and Aunt Sue at supper.

"Can't stop you, George," grunted Uncle Mose. "Wouldn't if I could."

Aunt Sue was close to crying. "You aren't leaving us too, are you, Jim?"

"No, ma'am."

So George left the farm, never more sure that he should. Still, his concern over where he might eat and sleep loomed large the first evening in Neosho as he loitered around the locked-up school. Of course Aunt Sue had packed some food for him, so eating was not an immediate worry. But sleeping was. He gawked anxiously about the area. And across a chest-high fence right close to the school was a sight for tired eyes.

"A stable!"

It was not large enough to stable more than a couple of horses, but it had a hayloft. George slipped inside — quietly but not so suspiciously that the one mule inside would start a ruckus — and scrambled up a ladder into the loft. There he

37

snuggled into the hay, feeling not much different than he had his entire life. After all he had slept on hay in a loft as long as he could remember. Of course he had always gone to sleep next to Jim, and often to the sweet sounds of a crackling fire. He had to remind himself several times he was not alone now. God was with him. At long last George was taking God's wonderful gift of one talent to school. He slept soundly, but the next morning he awoke with a start to the crowing of roosters. Brushing off hay, he scrambled out of the stable. Nearby was a pile of firewood. He sat on it and let the morning sun warm him.

"What are you doing there on our woodpile?" called a woman's voice from the nearest house.

"I'm waiting for school to open, ma'am," he answered.

Should he run? Would a bushwhacker come busting out of that house? George was startled to see a small colored woman emerge from the front porch of the house and come rushing toward him. But she was eyeing not him but the woodpile. Of course, he realized, she's getting wood for the morning fire. He jumped off the wood.

"Sorry, ma'am, I was just resting."

"Don't you know school doesn't open for another hour or so?" asked the woman. "Say, where are you from, youngster?"

"I'm from a farm over yonder in Diamond Grove."

"That explains it. What's your name?"

"I'm Carver's George."

"No, you're not! That much I know for sure."

The small, wiry woman enveloped George like a whirlwind. It seemed just seconds later that he was in her kitchen, building the breakfast fire for her, while she lectured how slavery was over. President Lincoln had freed four million colored folks in 1864 with the "Emancipation Proclamation." Including a certain youngster who was no longer "Carver's George" but "George Carver" — if he wished to be. She explained how the nearby school had been set up by the Freedmen's Bureau for colored children.

"Seems to me, youngster," she said as she bustled about the kitchen, "that God let you be born just about the right time!"

They were soon joined by the fiery woman's husband. Both the woman and her husband were very open with George. Andrew Watkins was a jack-of-all-trades fixer who was away from the house all day long. They had no children. The woman, Mariah, was a laundress and a seamstress. She was also a midwife, called far and wide to help deliver babies of every color. So, she stated, she could sure use some good help around the house. Before school started, George had been taken in by Mariah and Andrew. George could sleep in the loft in their house. They made it plain enough he would have to work, too, but George could tell "Aunt Mariah" was pleased with the way he scurried around the kitchen helping her. She didn't even have to ask him. "Uncle Andy" seemed just as pleased, as George cleared the table of the breakfast dishes and began washing them. George was so grateful for the biscuits and molasses he would have stayed and worked longer if Aunt Mariah had not shooed him off to school. Children were swarming into the split-log cabin that was about fifteen feet long on a side. At the door stood a colored man.

"I'm George Carver from over yonder in Diamond Grove, sir," explained George to the man.

"Well, I'm Stephen S. Frost," explained the man. "I'm the teacher here at Lincoln School."

"Yes, sir!" exclaimed George.

A colored teacher! That told George for sure he was headed in the right direction. Imagine. He might have been upset otherwise, because the school was crowded. By George's careful estimation there must have been seventy-five children jammed shoulder to shoulder. The crowding felt good at first, in the morning chill. But later in the day the heat from the closeness was deadening. At recess George saw Aunt Mariah right across the fence from school, in her own yard scrubbing clothes on a washboard immersed in a tub of sudsy water. Next to her were a basket heaped with dirty clothes and a basket half full of clothes she had already washed. She was amazed when he leaped the fence and began hanging the washed clothes on her clothes line.

"I've got fifteen minutes, ma'am," he explained. He sniffed, a big grin on his face. "Few things smell better than a garment dried in the sunshine and fresh air."

When the bell rang George leaped the fence and joined the other students. All the children clutched small rectangles of slate on which they scribbled letters and numbers with a piece of chalk. Stephen Frost's mission appeared to be to make sure every child could read and write. Numbers were used to make sure each child could at least count. But the children were not taught to add or subtract. Subjects like geography and history were rarely mentioned. But Stephen Frost did impart his personal philosophy.

"Know your place, children. No one is called 'Mister' or 'Missus' but white folks. Don't look white folks in the eye unless they ask you to. Always step out of their way. If you have a white friend who treats you nice, don't ever embarrass that friend by being familiar with him around other whites. . . ."

At first George accepted this as sound advice in a very dangerous world, but after a while it rankled him. George was not a big boy but he wanted to stand firm, not cower. Uncle Mose taught him a fellow is free. Uncle Mose taught him a fellow has a right to be different, too. Why did a colored person have to be so meek? Aunt Mariah insisted the Bible had all the answers. And George soon learned he didn't go only to Sunday school in Neosho. In Neosho he was also supposed to go to church at the African Methodist Episcopal Church.

He resisted at first. "No, ma'am. My Uncle Mose doesn't go to church. He says

42

a fellow doesn't have to go to church."

"Do you mean your uncle isn't a God-believing man?" she demanded angrily.

"He sure is a God-believing man, ma'am. But he's a Mason, and he says religion is the concern solely of the individual." George stood taller, just thinking about it.

"George Carver, I could just skin you!" shouted Aunt Mariah.

"Better come with us to church, George," said Uncle Andy firmly. On the way to church Uncle Andy whispered a rhyme, "She does believe without a doubt, that a Christian's got a right to shout."

Reverend Givens delivered the Word. George began to listen hard in church. He began reading Aunt Mariah's Bible, too. Yes, being meek was definitely the advice given by Jesus. Yes, turning the other cheek was also recommended. Yes, according to Jesus, anger itself was a very bad thing. And George began to understand that Jesus was telling folks how to throw off their chains of bitterness and resentment and hatred. Love your enemy. But George also learned the disciples themselves had difficulty at first accepting meekness. Especially Peter. The words in the Bible rang so true for George.

The advice of Jesus in one of the Gospels struck him forcefully:

Consider the lilies how they grow: they toil not, they spin not; and yet I say unto you, that Solomon in all his glory was not arrayed like one of these. If then God so clothe the grass, which is today in the field, and tomorrow is cast into the oven; how much more will he clothe you, O ye of little faith? And seek not ye what ye shall eat, or what ye shall drink, neither be ye of doubtful mind. For all these things do the nations of the world seek after: and your Father knoweth that ye have need of these things. But rather seek ye the kingdom of God; and all these things shall be added unto you.

So there it was: a promise from Jesus. George must not seek material things but seek the kingdom of God. The material things God would give him somehow. How George adored His advice. And meekness and even temper were important, too. George softened his opinion of Stephen Frost. Wasn't he echoing the words of Jesus? Did it matter if his meekness was not from his devotion to Jesus but from fear? Maybe it did. But George noticed too the respect white folks always showed Aunt Mariah. It seemed almost everyone in Neosho knew her. She too was meek and polite with white folks. He was sure her meekness was not from fear but from her love for Jesus. It was one more delicious puzzle to turn over and over in his mind, but sometimes he wished he had someone he could tell his deepest thoughts. One day someone did appear in Neosho to whom George could tell his deepest thoughts.

"Jim!" cried George.

Jim had finally come to Neosho to go to school, too. He was staying with the Wilson family. The brothers were reunited — for leisure times anyway. For life was not all school and unceasing toil. The Baynhams had given Jim a copy of *The Dime Ludicrous Speaker* and the brothers gave comic recitations from it. First Jim, in his fine tenor, with appropriate pomposity, would orate the story of "Where the hen scratches, there she expects to find the bug."

Then, hooking his thumbs behind imaginary suspenders, and drawing on every ounce of pathos, cynicism, sincerity, and irony, George would recite in his screechy soprano Mark Twain's "The Story of the Good Little Boy":

45

This good little boy read all the Sunday school books; they were his greatest delight. This was the whole secret of it. He believed in the good little boys they put in the Sunday school books . . . [He] had a noble ambition to be put in a Sunday school book. He wanted to be put in, with pictures representing him gloriously declining to lie to his mother, and her weeping for joy about it; and pictures representing him standing on the doorstep giving a penny to a poor beggar-woman with six children, and telling her to spend it freely, but not to be extravagant, because extravagance is a sin; and pictures of him magnanimously refusing to tell on the bad boy who always lay in wait for him around the corner as he came from school and welted him over the head with a lath . . .[1]

The brothers had great fun with their comic readings and learned many large words that they might never have encountered any other way. Sometimes now they created their own comedy. And it wasn't always comedy. Sometimes they debated whether women should vote and such troubling things. And sometimes they read books like *Uncle Tom's Cabin* or even *Little*

46

Women, and talked about them. They sang lively songs like "Jimmy Crack Corn" and "Old Dan Tucker." George played with other boys his age, too. Of course by now he was very handy with his pocketknife at mumbletypeg. And he could throw rocks like rifle shots. Back on the farm, before he knew any better, he had killed some birds that way. But killing sickened George, especially when he grew to love Jesus. Killing was a power he didn't want. He wanted to make things grow and live and multiply.

George shot marbles very well, too. Every boy had a small pouch of his "mibs," including one large shooter. George was no exception. But his hands were exceptional. His digits seemed twice too long and his oversized thumb would propel his "aggie" shooter like a cannon shot.

"I can't play 'keeps' with you anymore, George," every boy at school was soon saying after seeing George's aggie propel marble after marble out of the circle scratched in the dirt. "Or I won't have any mibs left."

But work beckoned George often. He improved his skills at this, too. From Aunt Mariah he learned to iron into the recesses of most garments. And to his crocheting he

47

added fine stitching, plaiting rag rugs, and making almost anything he could see. He had the knack of looking at a garment, not knowing how it was put together, and yet figuring out how to duplicate it. He could study a fancy lace collar in a store window and somehow he would knit an exact replica! Aunt Mariah said he had a special gift from God. But back on the farm, when Aunt Sue had once been able to keep from bursting into tears at the mention of his mother Mary's name, she had told George his dear mother had been just like that, too.

George took other jobs in Neosho. One time he lived in a house for a few days while the owners were gone back east to St. Louis. While he was there he had made many repairs — tightening the hinges on a screen door and such — and came home to Aunt Mariah as proud as a barnyard rooster. He reeled off a list of his many accomplishments. But she glared at him. "It doesn't matter how many things you did, but how well you did them," she insisted.

One day after George had lived in Neosho about a year Jim surprised him. "George, I'm not going back to school this fall. I've got me a good full-time job plastering walls. Besides, I'm not learning anything new at school. Are you?"

"Sure . . . ," said George unconvincingly.

The truth was that school was mainly getting the colored children to read and write and count. Stephen Frost never had time for any instruction beyond that. That fact weighed heavily on George, even though he still attended every day and clutched his slate, wedged in among the others. As winter approached he became very restless. School had become a waste of time, he had to admit reluctantly. Going back to the farm did not seem appealing either. Winter was a time of fixing things, not planting. George liked to fix things but he loved the planting more. He could never learn enough about that. But why was he thinking about going back to the farm? Wasn't that burying his one talent?

"The Smiths are moving on over to Fort Scott in Kansas," said Uncle Andy at supper one day.

"Kansas?" asked George with great interest. "Do they have a school for colored folks over there?"

"Reckon so," answered Uncle Andy.

But why should George try Fort Scott, seventy-five miles away, and not Joplin, just twenty miles away? It just seemed to him Joplin was another Neosho, a lead-mining town with a history of slaves. And white folks in this area still talked about Carthage and Pea Ridge with such stony regret and squinty eyes George just knew they were talking about Civil War battles. As often as not the speakers had been Confederate soldiers or, worst of all, bush-whackers. George had to admit the hard, hateful looks of old bushwhackers on the streets made him nervous, even when he dropped his eyes and gave them a wide berth like Stephen Frost told him to. So George started asking a lot of questions about Kansas and Fort Scott. He learned Kansas had always been a Free State, in spite of considerable pressure to join the Confederates. And Fort Scott, folks said, was where the legendary old abolitionist John Brown had been from time to time. So Kansas and Fort Scott seemed prom-ising and who knew how soon it would be again before George could hitch a ride all the way over there? But parting with the Carvers, the Watkins couple, and espe-cially Jim would be very hard.

Jim was shaken by George's restlessness. "What if Mother came back looking for us? She couldn't find us."

"Mother?" George was startled. Did Jim

still think of Mother a lot? "But colored folks have been free more than ten years now. Don't you think Mother would have come back by now?"

Jim couldn't answer that. "Say, George, let's get our picture taken," he said brightly. "There's a fellow downtown that takes photographs."

"But we can't afford . . ."

"Yes, we can. That way I'll always have a picture of you, brother, and you will always have a picture of me."

So the brothers sat for their portrait together, both dressed in their church clothes. They wore dark blue suits and clean, white ruffled shirts. George's tie was loose, hanging down over his lapel, but Jim had tied a neat bow for himself. George parted his hair on the right side; Jim parted his hair in the middle. But these differences were scarcely noticeable compared to the other differences in the two brothers. Although they were only four or so years apart in age, they appeared to be father and son. George was a boy, obviously in his early teens. Jim was a good-sized man, a head taller than George — even sitting down!

"Why, Jim's holding a book in his hand," noticed George later as he studied the pic-

51

ture. "I guess school meant a lot more to Jim than I realized."

George didn't fret over how he himself seemed to be fumbling with his awkward, oversized hands in the picture, because his heart ached so much at the thought of leaving Jim. But there was no turning back now. One day the Smiths pulled two large wagons up in front of the Watkins' house. They weren't covered with canvas tops but they were still Conestogas, the kind of wagons folks called prairie schooners. The Smiths were moving in the coldest part of winter as was the custom of the time. The muddy roads were frozen solid then, as were most of the creeks, so the wagons rolled hard but steadily ahead. Aunt Mariah, Uncle Andy, and Jim were there to see them off.

"Hold on there, George Carver!"

"Why, it's Stephen Frost," blurted out George.

"Here's a Certificate of Merit," said Stephen Frost breathlessly. "Show that to your next school teacher, George, so you don't waste time on things you already know."

"It's dated December 22, 1876," said George proudly.

Aunt Mariah presented George with a

Bible as he climbed onto the Smiths' wagon with his bundle of possessions. With the Bible he felt like his belongings had increased a thousandfold. There wasn't anything in the world he would have rather received as a gift. Now he had God's Word with him at all times. And he was sure going to need the warmth of the Word as, atop the wagon, the January wind gnashed him with icy teeth. . . .

four

The trip to Fort Scott was slow in the great jumbling, jangling wagons. And facing the icy north wind was foolish. So George, as sinuous and supple as an otter, burrowed down among the furniture. He crocheted and read his brand-new Bible. He no longer read it willy-nilly but tried to remember the specific chapter and verse of his favorite passages. As long as he read his Bible or kept busy he didn't fret. But if he was idle he began to worry. This was quite a dangerous thing he was doing, going off to Fort Scott as a boy of only thirteen or so.

Once in a while, when the wagons had to labor up a rise or a creek bank everyone had to get out and walk beside the great wheels. Then George drank in the countryside. The wonderful stands of timber had thinned out. And the rugged hills gave way to rolling plains of winter-drab grass. Timber retreated to the river valleys. More and more swatches of grass appeared. Winter had beaten the grass down, but

George could see the grass would stand as tall as a man's head in the summer. When they camped George collected the grasses and studied them. There seemed to be four kinds. Ground-nesting prairie chickens and quail would erupt underfoot and flutter off, low and hard.

"Fort Scott up ahead!" clamored the Smiths one morning.

"Are we in Kansas?" asked George.

Somehow he had expected to cross a river or something. But no. Suddenly they were in Kansas. Maybe there wasn't such a difference between Missouri and Kansas after all, worried George. Soon they were in Fort Scott, rattling north up Main Street. George did not want to go all the way to the downtown area, so he asked the Smiths to let him off. Farewell was short, and shouldering his bundle George headed west where he had seen some fine homes. It turned out these homes were on Jones Street. One Bible passage in the book of Matthew had struck him while he was reading during the journey:

Ask, and it shall be given you; seek, and ye shall find; knock, and it shall be opened unto you: For every one that asketh receiveth; and he that seeketh findeth; and to him that knocketh it shall be opened.

"Nothing to do now, God, but knock and ask," said George.

He walked up to a fine two-story, white-frame home, then virtually tiptoed to the back door and knocked. He took off his hat and clutched it in his hands in front of him. A white woman in an apron opened the door.

"Get away from here!" she snarled.

George dashed away. He couldn't recall seeing such a look of hatred in his life. Should he continue? What a foolish question! Of course he must continue. There was no warm kitchen with Aunt Mariah in it here. He must find a place. And quickly. January could be deadly for a boy, even hiding under the hay in a loft.

The next place was not as unfriendly. "What do you want?" asked the pale-faced woman who answered his knock on the back door.

"I'm looking for work, ma'am."

"We have none!" She slammed the door.

He almost had to laugh. Feeling more lighthearted he pretended Jim was with him. Yes, that reaction could not be classified as hatred. He would have to classify door-slamming as rude. It was many houses later, each one less imposing than the previous one, that George heard of a household looking for help. By early afternoon he had reached Wall Street and headed east until Scott Avenue. He seemed to be among stores and livery stables. But still he spotted a white-frame house, neat but not as imposing as the houses on Jones Street.

"Well, what is it?" asked a tall, severe white woman who responded to his knock.

"Ma'am, I'm George Carver from over yonder in Missouri. Ma'am, I heard you were looking for help."

"For a girl, yes. . . ." Her voice was exasperated.

George nervously wrung his hat. "Ma'am, I've been trained to do housework."

"That's possible, I suppose . . . ," she hesitated.

"Ma'am, I can dust and sweep and launder and iron the finest apparel. . . ."

"Look me in the eye!"

"Surely, ma'am."

"Mmmm, you appear to be just a boy. Well, step inside. It's freezing out. I'm Mrs. Felix Payne."

"Thank you, ma'am."

George stood in a small, cluttered kitchen. Mrs. Payne seemed confused, as if there was such a mountain of work to be done she didn't know where this George Carver from Missouri should start. Or was she trying to decide whether or not to throw him out?

"Well, I guess the dinner should come first," she muttered to herself. She turned to George. "You can start with all these dirty dishes. I don't suppose you can cook?" she asked, doubt causing the question to die off.

"Surely, ma'am, I've been trained to cook."

"You can cook?" Her eyes narrowed. "Mr. Payne is very particular."

"In that case if you'd tell me exactly how he likes his tucker, ma'am, I'd be sure to please him."

George could see Mrs. Payne was suspicious. Was this boy bluffing? "All right," she declared forcefully, "I might as well find out if you're the real thing or one of those Missouri windbags!"

"It will be a pleasure, ma'am," said George with real enthusiasm because, in spite of her small kitchen, Mrs. Payne had one of the new parlor stoves. He could hardly wait to stoke it up and get started.

Why, it would be a pleasure to cook on such a wonder — even if they threw him out by the scruff of the neck afterwards. "Perhaps, ma'am, you'd tell me exactly how Mr. Payne likes his tucker. . . ."

"You said that once before!" she said irritably. "Well, you didn't think I was just going to turn you loose in here, did you? Of course, I'm going to tell you how. And we don't call our meals tucker or grub, like ignorant backwoods folks over in Missouri. We call it dinner or lunch or breakfast, George."

Soon George was preparing what the Paynes considered standard winter fare: roast ham, pickled beets, mashed potatoes, buttermilk biscuits, milk gravy, and bread pudding. While he prepared dinner Mrs. Payne hovered over him, giving minute directions. It occurred to him more than once that she might as well have cooked it herself. But once George got used to Mrs. Payne's haughty ways — she was herself quite a windbag — he didn't mind working for her at all.

George heard Mr. Felix Payne remark after dinner, "The meal was excellent, my dear."

"The new boy George cooked it, dear."

"The boy who has been serving us

59

dinner also cooked it?" he said in astonishment. "It was so good I was sure you did it, dear."

"No, George did it. He does housework, too."

"Housework, too?"

"He's trained in it," she said proudly. "I suspect he's just about the only houseboy in Fort Scott with such training. . . ."

"Say, we better keep him right here. You have him put his gear in that spare room under the back steps, dear."

"That would be very wise, Felix."

George delivered on his promises. Mrs. Payne didn't have to tell him how to sweep and dust and make beds. He laundered and ironed with no instruction at all. Felix Payne seemed to be always shaking his head: "George ironed this shirt?" "George mended this sock?" "George cobbled this boot I've been having trouble with?" And fortunately where George did need instruction Mrs. Payne could not resist instruction. For she ruled her wonderful modern kitchen like a tyrant. "The stove is hot enough now, George." "A pinch more salt, George!" "Don't blanch the turnips too long, George!" "Now add a teaspoon of ginger, George."

Besides the Carvers, who treated George

like a son, the Paynes were the first white folks George had really dealt with on a personal basis. Although they became familiar with him he suspected there was a line he should never cross with them. And he remembered Stephen Frost's warning: don't ever be familiar with white friends when they are with other whites. So he was very cautious. He took no liberties with the Paynes. The book of Proverbs became his guide. There were hundreds of proverbs in the book and he read them all, over and over. Verses from Chapter 13 he wore like armor:

He that keepeth his mouth keepeth his life: but he that openeth wide his lips shall have destruction.

The soul of the sluggard desireth, and hath nothing: but the soul of the diligent shall be made fat.

A righteous man hateth lying: but a wicked man is loathsome, and cometh to shame.

Righteousness keepeth him that is upright in the way: but wickedness overthroweth the sinner.

Discretion, diligence, truthfulness, righteousness. In his heart George was living

61

the Word of God. Before one month had passed Mrs. Payne had him enter the baking contest of the Methodist Church. Within the building of soft native sandstone at Hickory and Jones, George's wares were tasted and judged. No one was prouder than Mrs. Payne when George won prizes for his bread and his buttermilk biscuits. "He's trained in the culinary arts, you know," she told everyone. She stayed close to him too, he suspected, so he wouldn't be stolen by another household. But George did not trek to Fort Scott to win baking contests. Nor did he come to Fort Scott to work for Mrs. Payne, nor to work for the Stadden Grocery Store across the street where he sometimes picked up some extra pay. He came to Fort Scott to continue his schooling. He saved every penny of his salary from his two jobs. He already knew Central School, a very fine three-story building of blond brick, was not for the likes of him. On one of his few afternoons off he found in the east part of Fort Scott, at Hickory and Margrave near the railroad tracks, a small soft sandstone building called East School. It was here in the din of steam-belching locomotives the colored scholars toiled.

"What!" exclaimed Mrs. Payne when

George told her he could no longer work for her.

"Ma'am, I'm sorry, but I've got to continue my schooling."

"But whatever for?"

"I've got to know the names of things, ma'am," he said weakly.

George rented a room in an old shack near the stagecoach depot. He bought his books, which now included geography and history, and spent every waking moment studying. His eyes were young and reading late at night by a flickering candle was no problem at all. When summer came he found a job at the Wilder House, a very prominent hotel at Main and Wall Streets. He laundered and ironed bedding. In time the word got out that George was very good at ironing men's dress shirts, too. George had really been near the center of town all the while, but he seemed isolated from it while working for Mrs. Payne. There were hotels and stores of every kind within shouting distance of her house. The Fort Scott *Monitor*. Penniman Hardware. Rothfuss Cigars. State Bank. Morley's Hardware. Liepman's Clothing. Robinson's Grocery. Prichard & Brothers Drugs. And the Wilder House.

George already knew the genteel talk of

the ladies in their church and garden circles. Now he learned what men in the teeming Fort Scott world of commerce talked about. Population was a very large topic of conversation. Fort Scott had forty-eight hundred residents and was skyrocketing upwards at an additional hundred per year. The great grasshopper plague of 1874 had slim pickings in Fort Scott, they recalled wryly. A cataclysmic fire had struck the downtown merchants the year before that. John Withers lost $2,000 of his $5,000 clothing stock. Glunz lost $10,000 in leather and harnesses. Randolph lost half his dry goods stock of $12,000. Tobacconist Dillard lost $2,000. Isett lost $5,000 in groceries. On and on the men talked.

"Money. Goods. Population. Money. Goods," echoed George in a whisper to himself.

But George heard stories even less appealing. For as long as George continued to work at something, he seemed invisible to white men. He might have been an ox treading grain for all they cared. It was from these aloof, self-important men he learned Fort Scott was still angry from the Civil War. Before the war there had been a hotel called the Free State Hotel.

Facing it was the Western Hotel, which welcomed the slavers. The town seethed with animosity over slavery. Border ruffians came across to burn and murder. Abolitionists like John Brown came there to persuade and plan. Now even after the war the men often took sides again.

"Colored folks have to walk very softly in Fort Scott," concluded George.

Women who didn't know their place were not welcomed by the men of Fort Scott, either. Susan Anthony, who came there occasionally to visit her brother Jacob Anthony, was outspoken for the women's right to vote. With glee the ladies invited her to speak at their church circles. There Susan Anthony lambasted her weak sisters for being submissive to the tyranny of males. The men loved to tell another story of Susan Anthony. Their hero was Ben Files, manager of the stagecoach line. It seemed Susan Anthony approached him to appeal for free passage in her role as a do-gooder. According to this male version, Files said coolly to the would-be mooch, "I'm sorry, Miss Anthony, this stage line carries no free passengers; it treats the rich and the poor, the prominent and the humble, alike." Then Files, ever the gentleman according to this whiskey and

65

cigar-smoke version, charged her six dollars for her fare, and when she complained about it he weighed her luggage and charged her more for excess weight! George didn't ponder that story very long. It had the smell of lies about it.

Once George rescued the *Monitor* newspaper from a wastebasket in a hotel room and read a poem written by the local poet who dubbed himself "Ironquill":

In a very humble cot,
In a rather quiet spot,
In the suds and in the soap,
Worked a woman full of hope;
Working, singing, all alone,
In a sort of undertone,
"With a Savior for a friend,
He will keep me to the end."

The "woman" could have been George himself! Was it possible the poet had passed by, thinking George, with his soprano voice, was a woman? He really had to laugh. George laughed at some stories he came across in downtown Fort Scott and trembled at others. Compared to women, men could be much more humorous — or much more violent. . . .

"What are you doing with them books,

boy?" demanded a white man on the street one day.

"These are my school books, sir," replied George, as stoically as possible, keeping his eyes on the ground.

If the man had been alone George might have escaped unscathed, but the man had a companion. The companion said, "Ain't no nigger who can afford books that nice. He musta stole them off decent folks."

It was just like Uncle Mose used to say: "Two scalawags do ten times the mischief of one." Before the hour was out George had been beaten foggy-headed by the two men and robbed of his books. He could have lied to them. He could have said he was delivering the books to Mr. Felix Payne. That probably would have stopped the ruffians. But George didn't want to lie. Being as "wise as serpents" did not allow lying. But he was stymied now. If he bought new books he would have no money for rent and food.

Somehow he didn't have the heart to work at the Wilder House again. He felt miserable. Even guilty, too. He hadn't always remembered his proverbs. He had been proud and told too many white folks of his plans. How the rowdy men would tease him now. So this upstart colored boy thought he was going to school, did he? Well, that didn't last long, did it? So George found work with a colored blacksmith, who didn't ask questions. The blacksmith had an invalid wife and was as busy working as George was.

One March day in 1879 the blacksmith took him aside, with a mortified look on his face. "Watch out for yourself, George. There's a colored man in the jail. He's accused of raping a twelve-year-old white girl!"

That crime against a child was so terrible it made George's skin crawl with dread. So he was not surprised at the wild talk that flamed around Fort Scott that afternoon. But when talk became whispers and hundreds of folks started gathering in the streets near the jail he became very nervous. He had heard of lynchings, but this was Fort Scott in Free Kansas, wasn't it? He crept near the jail to watch. The sun was low now, so it wasn't hard to keep out of sight. Reality exceeded his worst fears when about thirty masked men brandishing guns appeared among the crowd. George couldn't believe his eyes. Where was the sheriff? The crowd began chanting the filthiest oaths against colored people George had ever heard. He couldn't believe his ears now. He was numb with fear. George thanked God the sun had gone down. He backed well away into the darkness of the night. But he didn't run. He had to see what happened, even though now the distance and the night shrouded the events in mystery. Suddenly the ruckus increased ten-fold. Had the masked men stormed the jail?

"No! No!" screamed someone. "You've got the wrong man!"

Swept along amidst the crowd was a colored man, struggling. Yes, the mob had dragged him from jail! They were crazy with hate. George ran to the livery and hid. After a time the blacksmith came in and lit a lamp. His hands were shaking. His face sagged with sadness.

"What happened?" asked George, emerging from the shadows.

The blacksmith's mumblings couldn't convey a fraction of the intensity of what happened — but it was enough. The mob hooted and howled as the man was paraded down the street, then hanged. But that was not enough revenge for them. The dead man was drenched in oil and burned! George was sick with anger. What good did it do to go to school? What could he ever hope to accomplish among such

69

people? The blacksmith seemed to recover as he tried to calm George. Surely the anger of the mob was directed against the criminal for such a terrible crime against a child. They hadn't done it because he was colored.

"No, sir," replied George. "I heard many threats of the vilest sort against colored folks."

"But those were a few bushwhackers still bitter about the war," insisted the blacksmith feebly.

"No, sir, I'm leaving Fort Scott," said George. "It's too hard to tell bushwhackers from respectable folks here."

five

"I'm not running off like a befuddled calf either," muttered George as he left Fort Scott.

Ever since his humiliating beating on the street he had thought about such a trek. He had absorbed a lot of information in downtown Fort Scott. Even though the railroads hadn't been hiring men for the last five years it was best to stay close to the railroads. That was where folks were settling. First George had to head northwest to Emporia. From there he could follow the railroad in pretty much any direction. But he wanted to stay north and west. He had heard folks settling in these new areas were from Yankee places like Pennsylvania and Old Abe's Illinois. Maybe even from far-off lands, too.

If he eventually went west following the railroads he had two choices. The more northerly rail route, the Kansas Pacific line, went from Kansas City west all across Kansas and two hundred miles of Colorado to Denver. The more southerly rail

71

line was the Atchison, Topeka and Santa Fe that meandered more or less along the Arkansas River toward Colorado. But it went through Dodge City.

"Stay away from Dodge City," the colored blacksmith in Fort Scott had warned. "That's about the only place in Kansas where the Texans are dropping their cattle herds nowadays. A lot of Texans fought for the Confederacy and they're a mite touchy about free colored folks. When they start drinking whiskey they can get downright unfriendly."

Also unfriendly were outlaws, for these were the days of roaming outlaw bands. Jesse James and his gang had not been heard from since their failed Northfield, Minnesota, bank robbery in 1876, but the desperadoes were not dead. Folks knew that for sure. And it was a fact that the James gang often strayed into Kansas. There were plenty of lesser-known bad men, too, like Dutch Henry, who might shoot a wandering man to silence him or just to steal his grub. And lawmen like Bat Masterson and Wyatt Earp were quick to shoot and ask questions later. Still, George knew meeting Jesse James or Dutch Henry or Wyatt Earp was less likely than getting snakebit or getting struck by lightning.

Now those were accidents worth fretting over. Snakes were hard to see in the tall grass and a fellow sure felt like a lightning rod out in these treeless plains.

George first tromped into uplands called the Flint Hills. Here on these rolling uplands the four kinds of grass he had seen on the way to Fort Scott from Neosho were in full force. By now he knew they were called big bluestem, little bluestem, Indian grass, and switchgrass. Of course the buffalo that should have been there grazing on the grass were long gone. Every little town had the locals arguing over who shot the "last buffalo." But down in thicketed creek draws George saw plenty of white-tailed deer, the same kind he had seen all his life. They dressed out at about one hundred pounds of venison, but they were beyond his hunting ability. Prairie chickens and rabbits were not. George now carried a pocket full of rocks. Often enough he could stun one with rock. He hated killing but he had always been told a fellow had to have meat to survive. He gathered eggs of ground-nesting birds, too. Dozens of kinds of songbirds that fluttered through the woods of George's childhood didn't exist here. The prairie supported an abundance of birds but not much of a variety. He could recognize only the warbling meadowlark, a couple of kinds of sparrows, and a peculiar brown-headed grackle.

Once an Indian appeared, so suddenly he seemed a ghost. The red man also appeared to be wandering — as displaced as George and just as determined to live. He gestured at George's satchel. He seemed to be signing. Was he asking for food?

"I have no grub, sir," George explained.

"Grub!" the Indian snorted. "Maka ta omnicha!"

He slapped George on the shoulder roughly and motioned George to follow him into a thicket. There the red man found a small shrub and pulled it out by the roots. The roots had nodules on them that looked like beans. He rubbed the dirt off one nodule with his fingers and offered the cream-colored "bean" to George.

"Thank you, sir," gulped George, and popped it in his mouth. The nodule was deliciously rich and sweet. He would remember this shrub.

The Indian led George back into the open prairie. He found a tiny shrub and began digging energetically around its taproot. "Tipsin," he finally said, displaying how the taproot ballooned into what looked like a turnip.

Raw, tipsin tasted like a mild turnip. With much vigor the Indian made George understand this vegetable was the great staple of the prairie for red folks. Using the pot George carried in his gear he showed George how to boil maka ta omnicha and tipsin, too. The Indian also showed George how to find mendo, which tasted like a sweet potato. Another discovery tasted like a tomato. Although George already knew how to find and prepare various wild pickings like plums, chokecherries, elderberries, and onions, he politely listened. Often the Indian would make a gagging noise to show George eating a certain part of a plant was not only distasteful, but maybe deadly, too!

It seemed peculiar to meet an Indian in these times. The Indians had supposedly been harassed out of Kansas. No longer would a fellow run into Osage or Kansa or Pawnees in the east part of the state, it was claimed. Nor in the west would he any longer encounter Arapaho, Kiowa, Comanche, Cheyenne, or Sioux. It was these western tribes — the buffalo hunters — who most opposed the pioneers. But they had been subdued, the more powerful

75

tribes getting reservations, the weaker tribes getting nothing. The American bluecoats, hardened by the bloody Civil War, had little patience with the problems of Indians. When the seemingly invincible Sioux chief Crazy Horse was murdered up on a reservation in Nebraska in 1877, it appeared to be the death knell for the Indian way of life.

But thanks to this wandering Indian the doors of the prairie's grocery store had opened for George. George felt he could survive anywhere now as long as he could find water. And in the Flint Hills water was not a problem. One morning George awoke to find the Indian gone. For some reason he expected never to see the Indian again. When George reached Emporia he decided he would continue northwest to Junction City where he would come to the Kansas Pacific railroad. Then he could work the towns of the northern route, all the way east to Kansas City or all the way west to Denver.

South of Junction City in July George saw his first wheat farm. The crop was about waist high and shimmering like pure gold. The farmer, a gaunt man in his thirties, was cutting it with the handled kind of scythe folks called a cradle. Once in a while he would stop and sharpen the blade of the cradle with a whetstone. Behind the man walked two blond-headed boys, the larger one raking the fallen wheat into a pile with his hands, then bundling it into what were called sheaves. The younger one propped the bundles into stacks, a job called shocking. Sheaving was awkward for the older boy because he had to clutch the bundle while tying it with nothing more than stalks of wheat. George couldn't resist. He had to try it.

"Hey! What are you doing there?" yelled the farmer.

"Sir, I just wanted to get the hang of it."

"Well, I can't pay you anything."

George realized now that the boys were very young. One was no more than six or seven years old. He was flushed, sick-looking, with unblinking blue eyes. George had seen heat exhaustion before. He wanted to suggest the boy go rest inside a sod house that was not too far off. There was no other form of shade on this prairie. But telling a white man what to do was too dangerous.

"Sir, perhaps I can work for dinner," he suggested.

"Watch for snakes then," grunted the farmer. "Joey, you go inside the soddie."

George was relieved when the farmer agreed. He was even more relieved when the farmer sent the youngest boy over to a sod house. The foot-thick walls of a soddie kept it cool inside, even on a hot day like today. George had heard harvesting was the most dangerous time of the year on the prairie for snakes. Sometimes the surprised serpent was a nasty-tempered bull snake and sometimes it was the dreaded green-gray rattler. And the smaller the rattler's victim the more likely the victim would die.

"Dirty old snakes," complained the older boy, who now shocked as George sheaved.

George laughed. "If the Lord didn't have snakes guarding the wheat your pa wouldn't have one grain left to harvest for your bread. The reason the fields are crawling with snakes is that the fields are teeming with mice and rabbits that are only too happy to get fat on your pa's wheat."

Suddenly a rabbit burst forth. George automatically pulled a rock from his pocket and rifled it. The rabbit squealed and toppled over. George wrung its neck and left it for later. By the end of the afternoon he had dropped four rabbits. At the soddie later that evening he skinned the

78

rabbits and stewed them in his pot. In unbroken sod nearby he dug for vegetables like the Indian had taught him. These went into the stew, too.

The farmer was amazed. "Why, you've scraped together your own dinner."

But George had learned to eat when there was plenty. He followed his prairie stew with the dinner the farmer's wife had set out. He ate a chunk of ham, four fried eggs, a half a skillet of fried potatoes, and one loaf of cornbread. The farmer thought nothing of it. He ate the same way. If a man didn't eat that way when harvesting he would soon collapse.

"Why don't you stick around a few more days, George?" said the farmer after dinner.

"Sir, I think I will. I'd like to learn more about this wheat farming. All my folks farm corn."

The farmer was starved for an audience. He talked on and on about wheat. At first folks said he was stupid to stop raising sheep and sell all his good dogs except one and put all his hope in wheat, especially winter wheat. They said if he was determined to be a farmer, he could at least plant something practical like corn. But he heard of this wheat called "Turkey Red,"

79

brought into Kansas by the Russian Mennonites. There was a yarn about how a small Mennonite girl back in Russia had picked through thousands of wheat grains, meticulously removing only the red grains. These red grains when used for seed produced a miraculous variety of wheat that resisted drought and disease.

How George loved that story. He couldn't help quoting the Lord's words in Luke 10:21:

I thank thee, O Father, Lord of heaven and earth, that thou hast hid these things from the wise and prudent, and hast revealed them unto babes . . .

The farmer nodded and continued, "Seed cost me nearly thirty dollars to sow my twenty acres. Still, if a great many things don't go wrong we can harvest twenty bushels an acre. We could clear a couple of hundred dollars. Say, I'm not boring you with this talk, am I, George?"

"Sir, I like nothing better than good talk about plants and their uses."

"Some day maybe I can buy a real modern reaping machine," said the farmer dreamily. "A fellow just drives his mules through his wheat pulling that reaper.

When the wheat's cut it falls back on a canvas mat. My boys would be riding there on the machine binding the sheaves and throwing them into the field to be stacked later. We could cut and bind the entire twenty acres in four days with such a machine. As it is we'll be lucky to finish the job in three weeks. Right now I can't plant more because it takes too long to harvest. If you hadn't come along I probably wouldn't get my twenty acres done."

"Sir, you said a great many things could go wrong," reminded George.

Sensing George really did want to hear every word of wisdom he had on the subject the farmer talked on and on. Between the time the seeds were planted in October and the time the new wheat finished forming its own seed in July, nature could ruin the crop with blizzards and droughts and floods and wildfires and grasshoppers. A farmer was helpless to do anything about it.

"That sure is mighty frustrating, sir," offered George.

"Or it could just rain day after day until the grain rusts. A farmer can't cut the wheat when it's wet and the ground is muddy. Then it gets rusted so bad it's ruined. It's almost as painful as burying a stillborn baby when a farmer has to burn that tender wheat and plow the ashes back into the ground."

"That sure is mighty sad, sir," nodded George.

"And supposing your wheat is ripe like it is now," said the farmer. "Why, those clouds yonder could get that sick yellow-gray look and batter that tender wheat with a sky full of hail as big as wild plums. It's happened to folks plenty of times. Wheat can't be harvested at all then. It has to be burnt and plowed under."

"That sure is mighty aggravating, sir," commiserated George. "But why burn it?"

"If a farmer doesn't burn the wheat it will spring up again as 'volunteer' wheat. Volunteer wheat is no blessing. It's more trouble than it's worth because it isn't spread out evenly. Where it grows thin the weeds crowd it out. And where it grows thick the first little dry spell kills it."

"Sir, we're going to get your twenty acres of wheat in, for sure," said George. "I just know that's God's plan."

George excused himself, while there was still some daylight left to read his Bible. Reading the Bible was the only thing that could have disrupted his rapt listening. Plants and the way they grew fascinated him. There was a ramp behind the soddie and the farmer suggested he sleep on the roof. No point in letting too many critters share his bed. So George stretched out on the roof of the soddie and laid his head on his bundle of belongings. When the light was too dim to read his Bible he shifted his gaze to the heavens. To George the transfer was seamless.

Everything is God's creation, he thought.

The next morning George found his fingertips so sore he could scarcely touch anything. Binding sheaves of wheat had done that. But he forced himself to work anyway. He stayed with the farmer until the wheat was cut and shocked. Then he helped haul the sheaves in the wagon to the soddie. There the farmer rigged up a mule to a heavy roller weighted down with stone. The boys led the mule back and forth over the sheaves. Just for good measure George and the farmer, with handkerchiefs covering their nose and mouth, flailed the sheaves with willow branches. On the ground the sheaves broke down into a mix of precious grain and worthless chaff. That mixture was hauled load by load to the top of the soddie in a wheelbarrow and slowly poured over the side. The breeze winnowed the chaff away and the grain piled up by the soddie. When enough accumulated for a wagon load it was loaded and hauled off to the mill in Junction City.

"It's not the purest wheat I ever saw," said the farmer in a voice muffled by his handkerchief, "but it will sell!"

After the job of harvesting was done there was no time to rest for the farmer. He started busting more sod. As he plowed there was a constant popping sound. He explained to George that only virgin prairie popped like that, from roots folks called prairie shoestrings. George determined the roots to be the widespread, shallow roots of a small shrub. The shrub was called the leadplant because of its silvery gray leaves. Farming was fascinating, but George had no dreams of being a hired hand the rest of his life.

"I'm moving on, sir," said George.

"I got some good cash money for that wheat," said the farmer to George, "and I want you to have ten dollars."

"No, sir. I agreed to work for my grub." It made him unusually angry when folks tried to give him more than what he agreed to. It smacked of pity. Still, he liked this farmer. He tried to soften his voice, "Besides, sir, I enjoyed reading your books.

The Adventures of Tom Sawyer was very entertaining."

"Where are you headed?" asked the farmer.

"Maybe I'll go west."

"West!" The farmer was agitated. "Don't you know they call that wilderness the 'Great American Desert'?"

But that didn't deter George. Why was the government offering homesteads out west if the land was hopeless? So in September George went into Junction City. He learned that 1879 marked a rejuvenation in the railroads in Kansas. The previous five years had been very hard. Now they were building track again. George asked about a job.

"We're only sending crews out west now," said a railroad official, eyeing George. "The Little Wolf business is over."

"Little Wolf, sir?"

"It was that chief Wild Hog with Dull Knife who was behind the killing," blurted out another railroad man, trembling with anger. "They killed about forty settlers, mostly north of the Carlyle station."

George finally gleaned from the emotional outbursts that two chiefs, Little Wolf and Dull Knife, took three hundred Cheyenne off their reservation in Oklahoma territory and headed north late in 1878. Their journey across Kansas had been bloody. Telling tall tales was a favorite pastime on the prairie, so George always weighed every story carefully. But after a few days in Junction City he knew the story of the renegade Cheyennes was true. Dull Knife had surrendered, said his informants, but the fate of Little Wolf was unclear.

"I believe I'll be heading east toward Kansas City," he told himself. "I don't believe I care to meet Little Wolf just now."

So George drifted into northeastern Kansas. Eventually he heard the Cheyenne had been captured up in the Dakotas. But by that time he had found work in Olathe, just south of Kansas City. There he lived with Ben and Lucy Seymour. Lucy was a laundress, even more skilled than Aunt Mariah, and certainly far more skilled than George. From her he soon learned how to launder and iron the most delicate of ladies' garments. Ladies wore many fine undergarments that had to be as finely finished as the outer garments. So ironing them was a meticulous chore. But a fine party dress beat all in its demands. Sometimes it took even Aunt Lucy half a day to iron a beautiful dress with all its tucks and flutes and ruffles.

The first time George ironed such a magnificent dress of white organdy his heart was in his mouth. "One teeny scorch and it's ruined," he said aloud.

"You have the hands of a fancy fiddler," encouraged Aunt Lucy — but her eyes were wide with concern.

He went to school, too, finding one of the Freedmen's Bureau schools. And it was in Olathe that George got into regular churchgoing for the first time since Neosho. The Seymours were staunch Presbyterians. Unlike the African Methodist Episcopal Church in Neosho, which had a congregation of all colored people, the Presbyterian church was attended by only a handful of colored people. That was all right with George. The farmer near Junction City had almost restored George's trust in white folks. On days he managed to forget Fort Scott he felt very independent of his race. But he was surprised when Uncle Ben proposed he actually become member of the Presbyterian Church.

"How old are you, son?" asked Uncle Ben.

Normally George liked it when Ben called him "son," but the question of age always made George uncomfortable. Why hadn't he pinned Uncle Mose down? George hesitated. "Well, let me think on it a moment. . . ."

"What year were you born, son?" prompted Uncle Ben.

"Maybe it was summer of '64, because I heard the bushwhackers from Arkansas came back up in Missouri that winter. . . ."

"Why, that would make you not much more than fifteen years old!"

"On the other hand, maybe it was the summer of '61 because the Confederates just won the battles of Carthage and Wilson Creek, so the bushwhackers were feeling mighty frisky all that winter." He added brightly, "That would make me eighteen, wouldn't it?"

"Didn't anyone ever tell you, son?" asked Uncle Ben in exasperation.

"Why, no sir, they didn't. Uncle Mose's nephew Dick said once I was born near the end of the war, but Dick was a mite ornery, too."

Uncle Ben studied him. "You appear to be about sixteen to me. Why don't we just settle on '63?"

So George became a member of the Presbyterian Church. As far as he could tell it wasn't that much different from

Aunt Mariah's African Methodist Episcopal Church. The Presbyterians were governed by the Presbytery, a group of selected officials. The Methodists were governed by a body of bishops. Both churches offered two sacraments: baptism and the Lord's Supper. In both churches infants could be baptized. However, there did seem to be a difference of opinion between the two churches regarding predestination.

"Predestination . . . ," mulled George.

As near as he could tell Presbyterians said God predestined those to be saved; they were the "elect." They were saved no matter what. Methodists said no one was predestined to be saved but salvation was a personal choice for everyone. So was damnation, for that matter. There sure enough were passages in the Bible, like the greeting in 1 Peter 1:2, that spoke how God knew ahead of time who the elect were. Did that mean it didn't make any difference what a fellow did? Did that mean he was already either destined to be saved or to be damned? Finally George was able to resolve the difficulties in his own mind. God was outside of time, he reasoned. God knew the end of all things as easily as he knew the beginning. So in a sense he did know who was predestined to be saved. But that didn't mean those people had no free will to save themselves or damn themselves.

"We're moving, son," Uncle Ben told George one day in 1880.

"Where to, sir?"

"North-central Kansas, twenty miles or so north of Salina. The Kansas Pacific just laid a spur line up there and the town of Minneapolis is growing like wildfire. Why not come with us, son?"

"Certainly, sir," agreed George. Moving was rarely a problem to George, the drifter, and he was pleased to be asked. Still, this time there was a complication. "But I'd like to finish the school year first."

"You can join us up there later. You're always welcome to stay with us, son."

But things did not go smoothly after the Seymours left. . . .

six

George had to go down to Paola, a very small town on Bull Creek, to find work. Fortunately he found a school there willing to take him. Along the way he had acquired an accordion. It was the most frivolous thing he ever bought himself. It was enormous, with two bells he could ring while he wheezed out his tunes. He played by ear, trying to squeeze every note possible out of the serpentine instrument as he sang "Nearer My God To Thee" or "Home, Sweet Home." Once in Paola he performed for a music teacher. The man looked very sick and could only shake his head. Had George at long last found something he could not master?

"No, I just have to keep practicing," he told himself. "The right notes will come out of the accordion eventually."

But when summer came he had more important things to think about. He had finished the school year. Now he had to walk two hundred miles to Minneapolis! But he loved the outdoors and as long as

91

he had water he had no concerns at all. Determination could walk him forty to fifty miles a day. Just a few days after he began his journey he was following the newly constructed spur railway through the Solomon River Valley, walking through the burgs of Niles and Bennington. The land was not like the tall grass prairie he had seen in the east. He would have to find time to study it. One day at the end of the line was Minneapolis.

"George, you made it!" cried Aunt Lucy.

"Bring in your belongings, son," chimed in Uncle Ben.

The Seymours were living in a two-room shanty in a draw that ran through the middle of town. Lucy was nursing for a Doctor McHenry until she could establish her laundry business. Ben had found a farming job immediately and he hadn't the time to do much of anything to the house. The unpainted, vertical boards of the shanty were not chinked against the weather. In a flurry George began chinking. On Sunday he walked with the Seymours to the Presbyterian Church on the northwest corner of Third and Sheridan. It was a mere two minutes away. Not only was the service bracing, but George saw nearby a two-story building,

Minneapolis' only schoolhouse. Would they welcome a colored student?

"God willing I will start school when I get the time," he said.

But first George whitewashed the boards of the Seymours' dwelling and hung a sign that promised Fine Laundry. The demand for a fine laundress was great. The citizens tested Aunt Lucy, then laundry began flooding in from the well-to-do. Certainly many of the merchants and doctors needed shirts ironed as hard and shiny as porcelain. Their wives and daughters needed the very best attention on their party frocks. George helped her, but finally Uncle Ben urged him to take a job on a farm, too.

"There's good money to be made on the farms in the summer, son," said Uncle Ben.

So that was what George did. The farmers of Ottawa County were not like the farmers of Newton County in Missouri. None were from the south, except maybe Robert Bruce's wife, Barbara, who folks said hoofed it all the way from Georgia. She was an old-timer now, having braved the Kansas wilderness since 1863. Most of the farmers were Yankees from Pennsylvania like Peter Hanes or Christian Zuker, from New York like Henry

Jennings, from New Jersey like Fred Doering, or from Iowa like William Postlethwaite and Robert Bruce. And those who weren't Yankees were from far-off lands: Joseph Jagger, Tom Bradbury, John Allison, William Walmsley, Arthur Bishop, and John Lyne from England; David Mortimer from Scotland; Frederick Neaderhiser and Jost Hammerli from Switzerland; John Nelson from Ireland; Gustaf Hurtig from Sweden; and Adam Reh from Germany. Why, there were hundreds of folks in Ottawa County from England, over a hundred from Germany, even some from France!

"Imagine, folks here from France," George commented.

Most of the newcomers homesteaded but a few came and bought farms outright. Every one of them had his own ideas on how to make a farm. It didn't matter if the topic was as mundane as cisterns. It would be debated. Should the cistern be square or round? Should it be bricked or planked? Finally it would be tested in the unique Kansas earth, and often even then the debate would not be resolved. "Good heavens, you didn't plank it with cottonwood, did you? There's not a straight plank of cottonwood in Kansas!"

"Where did you get the sand for your mortar?"

"The local sand, he is no good!" the debater would counter with certainty in a thick German accent. "The mortar he holds him not together!"

George liked these farmers. They treated him like any other hired hand, no better and no worse. They respected hard work and honesty. A sense of humor helped, too. They talked of practical things that mattered but told some stories, too. Of course George had to learn ol' Tom Stewart shot the last buffalo in Ottawa County back in '70. But the tornado of 1879 was so recent it was high on the list of fireside anecdotes. Everyone had their own unique version. There was little humor in such a recent disaster. The wind was from the south the morning of May 30. About four o'clock that afternoon a funnel dropped out of a dark, troubled sky like an eggbeater from hell and scrambled a murderous path in the prairie for twelve miles. Many farmers lost buildings, but the Krones and the Vossmans on Salt Creek were hit the hardest. Krone's wife and daughter died — as well as a stranger who had run there for shelter. Vossman, too, lost his wife and daughter. Yet the Vossman baby boy was found in a field, plastered with mud but alive!

"Praise God for that," muttered George, but nature's wrath on poor innocents bothered him very much. Six died in all. The stranger seemed the saddest of all. His kinsfolk would never know what happened to him. How often could that fate have happened to George?

Of course no evening was complete without Indian stories. Had George heard about the Sioux attacks back in '68? The one in October was the worst. The Indians were very shrewd. They stayed away from Minneapolis. They divided into small bands and struck many farms at once. They killed John Andrews. Some farmers later found poor Alex Smith in the river, hugging a cottonwood but dead. Grandpa Smith died, and so did Peter Karns. Ulysses Seymore was shot full of arrows but lived. Mrs. Morgan and Miss White were captured by the raiders and taken away. The Indians burned many farms to the ground. The bluecoats didn't get the two ladies back until the next spring, both dressed like squaws. George Custer himself redeemed the ladies by jailing five Sioux chiefs until they were returned. Still, the Indians came back in June of '69.

Killed old Dyer and John Weir over by Indian Lookout.

"Have you seen the Vic Rees house in Minneapolis, George?" asked an excited storyteller. "He built it with loopholes for rifle barrels! Don't let your guard down, George. Indians were spotted roaming through here in '78. Folks say they're looking for horses to steal."

The third most-told story was the great grasshopper invasion of 1874. One sunny afternoon a great cloud bank roared toward the farms. Soon the cloud blocked the sun and delivered its load, not rain but grasshoppers! They poured down on everything, the pests much tinier than they usually are, but numbering in the billions. Hogs and chickens gobbled them up for a while but soon everything was overwhelmed by hoppers. The farmers lost field crops, gardens, trees, wood implements, clothing, saddles, and harnesses — just about everything that could be eaten! Some folks were smart enough to shovel dirt over their gardens and the buried potatoes came out all right.

"That next winter she was pretty rough," muttered one man in a thick Swedish accent.

On and on went the stories. Most were

true, but many were embellished. Drought, snakes, blizzards, Indians, wildfires, grasshoppers, tornadoes, wolves, floods. George had a few of his own to tell. Of course he didn't tell the tragedy of his mother and sister, but he didn't mind telling folks he had bummed around for a while in the tallgrass prairie with an Indian and he had harvested Turkey Red wheat. But George distinguished himself most of all by revealing his fancywork. Jaws dropped when he not only displayed his lacework but proved his quick hands could produce it.

"Und das George can cook strudel, too, I bet!" cried a German accent.

When George was given the job of hauling goods to and from town in a wagon he finally had an opportunity to really see the countryside. Timber was scarce, and what little was there was dominated by great cottonwood trees. Farming was best on the bottomlands along the river and creeks where the soil was a mixture of silt and clay rich in organic matter. Gullies revealed the soil was up to six feet thick. Where the sod had not been broken yet with a plow the virgin land was a blend of the tall grass prairie of the east and a short grass called buffalo grass. Buffalo grass was so hardy and so tough, he was told, that it was the only vegetation that survived the great grasshopper infestation of 1874. That sounded like a yarn but maybe it was true, reflected George.

"I surely would love to know," he said, always hungry for facts.

George now deciphered grasses in the peculiar way he had deciphered plants ever since he could remember, articulating many of his thoughts in a language he had invented. The waxy cabbage-colored bluestem grass was clumped, with jointed stems as tall as his head. From the stem the flat blades of the grass were up to two feet long and usually about three inches wide. The heads flowered in August and seeded out in September. There was a shorter variety of bluestem that looked greener to George's self-trained eyes. And its matted roots were markedly finer than the coarse roots of the tall variety. Was that why it dominated the uplands where moisture was more scarce? And what of the grayish-green buffalo grass? It grew to its tiny height of six inches only in the driest upland areas. Yes, it, too, had very fine roots. The seed heads were puny, making George suspect the grass spread mainly by runners. And in his trips he found other grasses: Indian grass, switchgrass, grama, and sideoats. There seemed a million things George could find out about these grasses if only he had time.

"Folks who think grasses are simple plants are looking at them but not seeing them," he concluded.

The variety of wildflowers and shrubs thrilled George, too. Sheep sorrel grew yellow, pink, and lavender flowers. Violets ranged from deep purple to almost white. The rainbow was there in blooms of primrose, larkspur, spiderwort, buttercup, vetch, verbena, aster, goldenrod, sunflower, petunias, pricklypear, and wind-flowers. On rare days he spotted the mallow flower and the Indian paintbrush. Birds were abundant. And he discovered their variety was a blend of the kinds he had seen in the east and some new very edible kinds like the prairie chicken and a plump bird that looked like quail. There were meadowlarks, doves, and horned larks, too.

But George couldn't dawdle too long among the birds and plants. His boss expected him to haul goods in the wagon. Oxen supplied the power. Oxen were still preferred by many farmers because they were cheaper than a span of horses — a fit pair of oxen could be bought for sixty dollars — and oxen could live off the pasture better. Their strength was immense; a quick flip of one great head while harnessing could send George flying. Fortunately, he reflected, the massive beasts were docile.

One day on a trip hauling fodder to town he called, "Gee!"

The oxen veered off to the right. There was one site he could no longer resist. He was detouring this one time to the place called the "city of rocks." Folks described it as a mystery of colossal proportions on the prairie. And finally he saw them. Golden brown boulders twenty feet in diameter were sitting on the prairie like giant marbles on a plate. But George didn't have much time to marvel at the boulders because water was nearby. The oxen bolted. "Gee! Haw!" he screamed. *Turn right, turn left! Do anything but don't careen straight ahead!* his mind screamed. But the oxen were thirsty. He prayed hard now as the oxen dragged the wagon farther and farther into the murky water. Finally the oxen stopped, slurped water for a while, then lurched back onto land, almost overturning the wagon.

"That's my last detour for a spell," gulped George.

101

That fall George went back to town to live with the Seymours. Strolling Main Street in just the one block from Mill Street to Ottawa Street he saw a post office, a courthouse, Joslin's Hardware, Barnes' General store, the Commercial Hotel, Heald's Grocery Store, Trow's Shoe Repair, Crosby's Drugstore, Slocum Barber Shop, a livery, and several other stores! There were just as many stores in the next block. Yes, Minneapolis was growing, prospering. A print shop, a wagon maker, a lumberyard, a mill, a cafe, a photographer, a millinery, a harness store, a jewelry store, a land office, a furniture store, two doctors' offices. Did Minneapolis lack anything?

To his relief he found town life as congenial as farm life. Oh, he heard Aunt Lucy referred to by some as that "old darky down in Poverty Gulch." He supposed he, too, was called "darky" and maybe worse. But aren't there always a few bad eggs in a town? He had missed the annual County Fair. Folks said it went on for four days east of town, highlighted by horse races. George was more interested in the exhibits and he would surely have entered his buttermilk biscuits. But there would be other fairs and he did not miss the spelling bees and ciphering bees later in the year, where he distinguished himself. Nor did he miss the husking bees or house-raising shindigs. And he began to play his accordion in trios and quartets and sing. Often he donned a long-tailed coat given him by a reverend and marched in parades, playing his accordion while strutting like a majorette. He began putting a flower in his coat lapel every day, which astonished folks no end in January!

He sang in the church choir, too. He even took parts in plays. His squeaky voice brought laughter from an audience, and often he would play a flighty woman. In "Poverty Gulch" Aunt Lucy had a thriving business, taking in more laundry than she could handle. George was only too happy to help when he was not in school, because, to his great relief, he found no opposition to his attending school. Officially classed a fifth-grader, he was the only colored student in the school. Almost any evening he could be found studying by a flickering candlelight. Getting a certificate for eight years was well within his reach now. But townsfolk were now talking about adding "high school," schooling that went beyond eight years. Well, he might as well do that, too.

Who knew where it might all end?

"Say, George," said Aunt Lucy one day, "someone told me there's another George Carver in town now."

"Another George Carver, ma'am?" he gasped. "I didn't know there was another George Carver in the whole wide world. Do you suppose he's getting my mail? I haven't heard from brother Jim in a good long spell."

"Add a middle initial, son," suggested Uncle Ben. "Haven't you noticed the signs all around? C. H. Hoyt? A. F. Shepard? C. C. Olney?"

"You're right, sir . . . ," mulled George, then exploded, "I like the sound of double-u."

"George W. Carver . . . ," tested Aunt Lucy. "Yes, that's nice sounding. What's the double-u stand for, George?"

"Stand for, ma'am? Does it have to stand for something? I'll have to think on that."

George should have taken a stand on the double-u because folks started calling him George Washington Carver. George Washington was a very popular combination of given names. George Washington This and George Washington That. But George found it embarrassing. George Washington, the father of America? No, he was just George W. Carver. No, actually he was just George Carver. But his discomfort did not end. Soon everyone was calling him George Washington Carver. And he had to admit in his less humble moments that the trio of names — six thumping syllables — sounded mighty impressive. But most of the time he wished he hadn't done it. And the mail he might have got because of it almost broke his heart.

"Oh, Lord, no!" he cried one day, after opening a letter from Aunt Mariah Watkins.

"What is it?" asked Aunt Lucy, her voice trembling.

"Jim is dead."

Numb, George pulled out his precious photograph of himself and Jim. Jim, his great strapping brother, had died of smallpox. He was buried in Seneca, Missouri. It was the greatest loss of George's life. And it magnified his other losses. He had lost his sister, whom he rarely thought about. He had lost his mother, whom he thought about only in the fuzziest forms. There were even two infants buried on the Carver farm that his mother had lost. There was the loss of his father, killed in a farming accident before George had been

105

born. It shamed George not to know anything at all about them. Oh Lord, had he ever asked Aunt Sue about them? George was inconsolable. What a sinner he was. He needed his Bible more than ever to try to understand such terrible losses and his own callous indifference. Oh, poor poor Jim.

But George continued his schooling. Thanks to Doctor McHenry he read by candlelight not only school books but the very latest books of all kinds. By 1882 George negotiated to buy his own lot from a local businessman. Aunt Lucy's laundry business was more than she could handle. She fully approved of George starting his own laundry. George's deed was made out to "George Carver." It had been one of those days George felt like he was flaunting "Washington" and couldn't bring himself to even include the double-u. Maybe he was distracted by the cost. The lot cost George Carver the staggering sum of one hundred dollars. He was to pay it off at five dollars a month for two years.

"But what if I miss a payment, sir?" he had asked the mortgage holder, somewhat worried. After all, he had school expenses.

"Don't worry, George. You'll catch up

on your payments," the businessman reassured him.

George was still in "Poverty Gulch." He had a one-room cabin with a lean-to kitchen behind. To ease access to his laundry George arranged planks and rocks across chronic mud holes. As if he didn't have enough to do he confided to friends he was thinking of writing a book entitled *Step by Step on the Golden Ladder.* He was sure his book would inspire other colored people to become educated. The tolerance in Minneapolis seemed to free him to become truly aware for the first time that he might have a mission for his people. But the more he discovered about the difficulties of getting a book of any kind published the more he regretted revealing this ambition, which even appeared in the local newspaper December 22, 1883.

"No harm done expressing my true sentiments, I guess," he told some, "but there are many other things I must do first."

He maintained his interest in plants, supplementing his diet with bounty from a garden. Yet often he bought specimens and paraphernalia he could ill afford. But how could he resist trying every new technique that reached his ears? Would a mixture of coal ashes, sulfur, and hellebore applied to his cabbages in the morning dew destroy pests? George had to know. He could scarcely sleep until he found out. And did adding linseed oil to his usual mixture of resin and beeswax make a better grafting wax? George had to know. He would rather go without food than not buy the linseed oil to find out. Sometimes he justified expenditures by applying his plant experiments to his laundry business.

"Can I isolate my own oxalic acid from a plant?" he wondered.

If so, he could mix it with the blue he made from copperas and prussiate of potash to make his own bluing. Indeed, sometimes he felt like a chemist dealing with laundry problems. Many problems required more than just judging accurately the amount of bluing or soap flakes or borax or ammonia or bleaching powder. Sal soda, unslacked lime, alcohol, glycerin, turpentine, carbonate of ammonia, alum, beeswax, salt, gum arabic, lemon juice, rice water, and chloride of lime were all used by George for dozens of different problems. Folks were loathe to throw clothing away. Occasionally, instead of removing a stain or yellowing from a fabric, he had to add color to the fabric. For this he learned to mix dyes, an exciting prospect since any dye left over he was sure to swab onto paper to create a flower. Brown he produced by mixing alum and catechu, yellow with bichromate of potash. Red triumphed from a complex chain of sumac, muriate of tin, burr wood, and oil of vitriol. Many of the solutions were suggested by George's customers, who came to Ottawa County not just from all over America but from all over the world. Often he modified; often he improvised.

But George's facility at solving problems often cost him more money than it earned. He was faced with missing payments once in a while or dropping out of school. He was so close to completing his eight grades the last option was unthinkable. Nothing on earth would keep him from graduating. And after he achieved that how could he refrain from taking a handful of what were called high school courses? After all, he was an esteemed student. His teacher Helen Eaker was so impressed by George that whenever he read an essay on plants or such she had all the other students in the school come and listen to him. She could do that because she was also the school principal.

One day the mortgage holder came to see him. "George, I have an empty lot over on the south side of town. Why don't we trade?"

"This location is very convenient to my customers, sir," answered George nervously.

"I'll even move this shack . . . I mean this house over to the other lot for you."

"I don't believe I would much care to move over yonder, sir."

"Your deed is void if you miss payments. And you have missed several payments."

"But I'll catch up with the payments, sir."

"I'll have some men come and move your shack . . . I mean your house . . . over to the other lot. I'll overlook the fact that you've missed payments."

"But, sir . . ."

seven

Anger boiled inside George, but he had learned long ago anger led to foolish actions. Proverbs 14:17 in the Bible confirmed it: "He that is soon angry dealeth foolishly." So he calmed himself, then thought long and hard. He had many friends in Minneapolis. He could have caused his mortgage holder a lot of trouble by enlisting sympathy for himself. But it was not his way to be a crybaby. He had learned a hard lesson. He had signed a loan agreement. The man had the letter of the law on his side. So George moved to the location on the south side of town as his oppressor wished. It was only later he heard that because Minneapolis was expanding so fast his old lot was now worth several hundred dollars. Oh, what he could have done with the money? Why, who knew? Maybe he could have gone on to college!

"But perhaps college is impossible anyway," he murmured.

Suddenly Philippians 4:13 hit him with force: "I can do all things through Christ which strengtheneth me." Did George believe God's Word or not? So he began to ask friends like Helen Eaker and Doctor McHenry if college was a possibility. At first they were hesitant but as they thought more on it they became enthusiastic. Why shouldn't George go to college? But where? Just who mentioned a small college in northeast Kansas called Highland College he couldn't remember. But somehow this college became his focus. The initial step was getting accepted. So George sent his school records and several letters of recommendation to Highland College.

"Oh please, God, let them accept me!" he prayed.

George waited with increasing anticipation. If he continued to save money and sold his current lot, even after paying off the greedy mortgage holder, he just might have enough money to get a good start in college. Once started there he would get a job on the side. He could make do. His head was swimming. What courses would he take? He loved painting and art. He loved plants. Oh, what a glorious choice he had to make — if he could get into college. "Oh God," he prayed, "please let me go to college." And several weeks later, George received his reply. . . .

"Yes!" he screamed. "It says right here I'm to report for matriculation on September 20, 1885."

George said his good-byes to his dear friends, the Seymours, and many others. Then he took the train to Kansas City where Chester Rarig, an old friend from Minneapolis, had started a business school. George wanted to learn shorthand and typing. The typewriter was a newfangled machine that could hammer out a word a second, Chester claimed. With shorthand and a typewriter George could compile volumes of college notes in the most efficient way. But he was very cautious in Kansas City. Cities and larger towns frightened him. There seemed a greater chance of running into crazies like the ones in Fort Scott.

Maybe he was wrong, he knew, but he just felt safer in smaller towns. That was why he had worked in Olathe in 1880 and not in Kansas City. And his situation was even more tenuous now. At twenty-two, George was a man. A stout man like his brother Jim he was not, but nevertheless he was too tall at six-foot and too dark as a grandchild of Africa to trigger the automatic amused tolerance from white folks his frail boyish appearance had triggered in

113

the past. He had to be more careful than ever. His suspicions of Kansas City were soon confirmed. One morning at a cafe with Chester Rarig, George was refused service.

"You can't do this!" complained Chester.

"Please, Chester, come outside with me," intervened George.

Outside he convinced Chester to go back in and enjoy his breakfast. He would be very unhappy if Chester did not to do that. George insisted he didn't want his friendship to cause Chester problems. Chester went back inside, shaking his head. But that was the way George really did feel. If Chester had not gone back inside he would have felt even worse. After that incident George became even more cautious. If he saw Chester with white folks George avoided him so he didn't risk Chester being embarrassed. Or if Chester could not be avoided George was polite but as stoic as stone.

But in spite of the occasional racial incidents George was soaring. Minneapolis was booming. His property there, on the south side of the town, had been sold for five hundred dollars! He wore a very nice checked gray suit now. He acquired his own typewriter. After becoming competent in shorthand and typing in just a few weeks, he stopped in Olathe, then Paola, to visit old friends. They were delighted with his prosperity. He went on to Missouri to stop in Diamond Grove and Neosho. Both the Carvers and the Watkinses could not have been more pleased for him if they had been his real parents. He visited Jim's grave. Uncle Mose had set a stone there.

"Why, Jim was born October 10, 1859!" exclaimed George. "Uncle Mose knew the very exact day."

Why hadn't he remembered to ask Uncle Mose about his own birth date? It was too late now. Soon George was on the train headed north out of Neosho to Highland College. The town of Highland was about as far northeast as one could go in Kansas without walking off a bluff into the Missouri River. The area was hilly with woods of oak and hickory, almost like his old woods he loved in Missouri. And yet as he walked to the college an herb with pink and lavender flowers drew him aside. It was the skeleton plant, a conspicuous prairie plant. Its presence in this woodland was a delicious mystery. Yes, George was going to be very happy here at

Highland College. He could imagine himself sampling every specimen. Thus excited, on September 20 he reported to the dean.

"You're colored!" exclaimed the dean.

"Yes, sir, that's true," agreed George.

"But we don't allow colored boys to matriculate here. . . ."

George was reeling. After the stinging pain stopped he felt like such a fool. He had spent almost his last penny acquiring clothes, a typewriter, train fares. Oh, foolish pride. He knew now he had strutted, not outwardly maybe but for sure in his heart. Yes, he had gone back in a new suit so one and all could see how George W. Carver had prospered! Maybe he had remained soft-spoken and acted humble, but inside he was choking on his pride. Still, he was devastated. This was the hardest blow since Jim's death. The business over his property in Minneapolis paled in comparison. But perhaps this was God's way of humbling him, as if a colored man needed to be humbled. But what was George thinking? Surely God was color-blind. Did George want special treatment from God because he was colored? He had been proud. Oh yes, how proud he had been.

George found no solace in his Bible this time.

He read one stinging reprimand after another. The Book of Proverbs alone made him feel as small as a grasshopper. "When pride cometh, then cometh shame," warned Proverbs 11:2. "Only by pride cometh contention," promised Proverbs 13:10. "Before destruction the heart of man is haughty," cautioned 18:12. "Let another man praise thee, and not thine own mouth; a stranger, and not thine own lips," said 27:2.

" 'A man's pride shall bring him low,' " read George from Proverbs 29:23.

In a moment of foolishness he wondered if this sin of pride was only peculiar to the Book of Proverbs — but no. The Old Testament resounded with its condemnation. "Thine eyes are upon the haughty, that thou mayest bring them down," stated the ancient 2 Samuel 22:28. Psalms 138:6 read, "Though the Lord be high, yet hath he respect unto the lowly: but the proud he knoweth afar off." No, the Old Testament offered no comfort to the proud.

The New Testament? George had only to read the words of the Lord in Luke 14:11: "For whosoever exalteth himself shall be abased; and he that humbleth himself shall be exalted." Yes, God had surely slapped him down for being so proud. God had not intended that he go to Highland College. George forgave Highland College, although they were more wrong than he was. He now had to earn a living again. He had a nice suit, an accordion, and a type-writer, but no food. Was his fate in life to just survive? To be a curiosity — George Carver, the colored man who crochets and expounds on every subject? Surely God intended him to do more than that. Again he sought guidance from the Bible. "In all thy ways acknowledge him, and he shall direct thy paths," said Proverbs 3:6. So George sought refuge in the church. "God, please guide me," he prayed.

"Are you George Carver?" brashly asked a woman of about fifty after the church service.

"Yes, ma'am, I am," he answered.

"Well, I heard that. . . ." She abruptly stopped, perhaps having read the pain in George's face. Or was it the dignity? "My husband John Beeler runs an orchard south of town," she finally said. "Have you done farm work?"

"All my life, ma'am. Housework, too, including laundering and ironing."

"Housework? Laundering? Ironing?

Then why should I waste you picking apples? You needn't tell my husband you've ever done farm work."

George was taken into their house, where predictably he impressed them with his efforts at housekeeping. No one was happier than their eighteen-year-old daughter Della, who despised housework. She was much happier playing Chopin on the piano. Nor was she pleased cooking for her father when Mrs. Beeler went on a trip.

"Father is impossible to please," she complained to George. "His false teeth are bad and he won't wear them."

"Perhaps I could help, ma'am," offered George.

"Now don't tell me you cook, too?" she asked suspiciously.

"If you would rather do it yourself, ma'am . . ."

"No! Please cook for him, George."

That evening John Beeler's eyes widened when he saw George's generous offering of mashed potatoes and milk gravy. He eagerly downed it. When he sampled George's cornbread he never came up for air until the entire pan was devoured. Only then did he try to reassure himself George would not be leaving any time soon. George was not anxious to stay. He wanted

119

to move on. But where? And what would he do? At any rate it was a bad idea to be adrift in Kansas in the winter time.

"Sir, I'll be staying here until next spring if it's all right with you."

John Beeler grinned a toothless smile. "Could you fix this same grub tomorrow evening, George?"

All winter George cogitated over what to do next. Not getting into college had crushed him. College was no longer an option. But he had no desire to forever launder and iron shirts. Nor did he want to be a handyman, a jack-of-all-trades. One of the happiest men George had ever met was the wheat farmer he had helped back near Junction City. That man not only was raising a nice family but he had a plan. So many bushels of seed to sow, so many bushels of wheat to reap and bind and thresh.

"George, I've been wondering," said John Beeler one day, "why don't you homestead?"

"All the open land is way out west, sir."

"There's land available out in Ness County."

"I don't recall a Ness County, sir."

"Well, it's off the beaten track, I'll admit. The Kansas Pacific runs a fair piece north

120

of Ness County. And the Atchison, Topeka and Santa Fe runs a fair piece south of Ness County."

"What's the land like, sir?" asked George, but he already had a fair idea: the Great American Desert.

"I'll let my son Frank tell you whenever he gets back to visit. He's been helping his uncles and cousins start up a town out there in Ness County. Right near Walnut Creek. Elmer Beeler has started a post office there and laid out a town. He's built a twenty-five-by-forty-foot stone building on Main Street. Bolivar Beeler has built a livery and a sod house hotel called the 'Beeler House.' And my son Frank runs the general store."

"What do they call their town, sir?"

"Beeler," shrugged John Beeler.

"I'll think on it, sir."

George could cipher very well. And that spring his ciphering almost got him so he couldn't sleep. Because in his mind he cleared and sowed ten acres of his homestead in Red Turkey wheat and reaped twenty bushels per acre. Wheat was selling at fifty cents a bushel, so George sold his crop for one hundred dollars. Of course, that was not free and clear but still, for a launderer and housekeeper, it was a tidy little sum. But why clear only ten acres? Why not forty? George's plans expanded every night. By summer George was clearing eighty acres. And why only twenty bushels per acre? Did anyone have a greener thumb than George? Why not thirty bushels of wheat per acre? His crop now yielded a staggering twelve-hundred dollars!

"Is it any wonder folks risk everything to go west and homestead!" he exclaimed restlessly.

It wasn't that George loved money. But think of the greenhouse he could build with his savings. Surrounded by acres of nurseries. When George met Frank Beeler he became even more enthusiastic. Frank was about George's age and bursting with energy and optimism. By August of 1886 George accompanied Frank when he returned to Beeler. They rode the Kansas Pacific train from Kansas City to the Wakeeney station. From there the two buckboarded south across forty miles of rolling, treeless prairie. Frank was quick to explain Beeler was getting a regular stage run in October. It was to be a regular old rockaway coach that just smacked of older times folks were already calling the Wild West. Of course the greatest boon to

Beeler would be the arrival of the railroad in the spring of 1887. Frank was excited. Then settlers would really flood in to Beeler. The population of the township was about two hundred. By next year it would double for sure. On the way across the short grass prairie George was also receiving a detailed summary of the agricultural situation. The local homesteaders were planting corn.

"But why not that good Russian wheat they grow back near Junction City?" asked George.

"Turkey Red?" Frank Beeler laughed. "Because the grain is hard as rock, George. Too hard for the mills they have around here. How can you raise a kind of wheat that can't be taken to a mill? A few have tried some soft-grained kinds of wheat here but they mostly die. They can't survive the winter. No, George, this is corn country — or nothing."

"But I should have sowed my corn last April or May. I thought I was coming to sow winter wheat. What am I going to do?"

"File for your homestead. Build a soddie on it, so you can demonstrate you're improving the land. Then hire out to someone. Save your money to buy the goods you need."

"What will I need?" asked George politely, although he had already thought on it a great deal.

"The few materials for the soddie will cost you about ten dollars. And if you cook like I do you'll need to buy a number eight cookstove, an iron pot or two, a griddle, a tea kettle, and a tin boiler. Whole shebang will cost you twenty-five dollars at my general store."

The burg of Beeler was about how George imagined it. Bolivar Beeler's sod house hotel called the Beeler House was an imposing sight. Such an enormous, sprawling soddie! Frank's wood-frame general store certainly lent a civilized touch. But to George the most imposing sight was Elmer Beeler's great yellow limestone building. The two-story edifice dominated Main Street!

But Elmer was not in a good mood. "That rat Dan Rineley sold the Atchison, Topeka and Santa Fe eighty acres of land south of here!"

Frank was stunned. "Do you mean the railroad isn't coming here?"

"Not to Beeler. It's going to their acreage south of here!"

George learned that the town merchants were already buying plots in the eighty acres of land the railroad owned. Bolivar was already planning to sell his great soddie and build an even greater stone hotel on the railroad acreage. Frank could certainly drag his frame store down there. But Elmer, the real founder of Beeler, could not move his stone building. He was seething.

"I best find George here some land," said Frank uncomfortably.

"George, if you need some work in the meantime there will be plenty of work to do at the new Beeler House," said Bolivar, and added cheerily, "We're calling the new town Beeler, too."

The first thing George did was scout out the land that was still available for homesteading. He had never been so excited. Imagine, George Carver owning 160 acres, plant life and all! He merely had to live on it five years and it was all his. What could stop him? He could virtually live off the land like an Indian if he had to. He studied the bottom lands along Walnut Creek. From there the land rose sharply about one hundred feet to level off into a tableland of gently undulating prairie. George felt more comfortable getting close to water. Springs were rare, according to Frank. And what little timber there was grew along the creeks. Of the native trees of hackberry, cottonwood, and box elder, only box elder was even marginally suitable for building. But maybe George could find an ash tree. Ash was fine wood.

"Just one ash tree would do nicely, Lord," he prayed.

The Beelers had certainly been right about the abundance of available land. Many sites were available. But they were all on the uplands. The choice bottomland sites with what little timber there was had been taken. So one mile south of Beeler George found a 160-acre site as close to what he wanted as possible. There wasn't a tree on it. Officially the property was the southeast quarter of Section 4 in Township 19 South, Range 26 West. From its north boundary Walnut Creek was half a mile north. From its west boundary Darr Creek was one-eighth of a mile west. He filed on it and, remembering his previous discussions about 1863 possibly being his birth year, he gave his age as twenty-three. He would have to dig a well on his new land.

"George, I loan out an auger from my store," said Frank. "That way you can sink test holes for water. Finding a good shallow spot for water will save you a lot of trouble. And of course you want your well dug before you build a permanent house."

Water was usually found before a depth of twenty-five feet, according to Frank. George knew that was not dangerously deep for a water well. But it made his skin crawl to think of actually climbing down that deep in the earth to dig. Still, many farmers did it. How George would have loved to throw up a stunning gold-tinted limestone house next to his water well, but the limestone was quarried thirty miles away near the burg of Bazine, according to Frank. Hauling even rough-cut stones to Beeler was quite expensive.

"There's some local lime, George," commiserated Frank, "but it's only good for plastering." He studied the late summer sky. "We could get an early frost out here, George. We better get started on your soddie."

"Frank, I'm just about out of money. I better get some work. How about Bolivar's hotel?"

Frank laughed. "Hire out harvesting sorghum first. The work only lasts a few weeks but it pays better. You'll find there's more work to do around here than workers, George. Just make sure your boss has the money to pay you." He looked at George's boots. "And take your boots off

127

whenever you can. This blamed buffalo grass just chews up leather."

So George split his time harvesting sorghum and building his soddie. Sorghum grew well over a man's head. The men first went through the field "stripping," ripping the leaves off the towering canes with a hay fork. Later they would go through the field, bending the cane over and lopping off the great seedy head with a huge knife. Then they severed the cane at the bottom. Some farmers rigged up their own crude presses from logs but most hauled the cane to a sorghum mill where it was first ground up. That mass was then pressed to squeeze the sap into vats. Then the sap was boiled and skimmed to make molasses.

When George wasn't working in sorghum fields he was tending his homestead. The native grassland had two kinds of grass, both of which George knew from Minneapolis: buffalo grass and blue grama. George was now convinced the relatively simple vegetation of the shortgrass prairie allowed only a few kinds of birds. Prairie chickens, a quail-like bird, doves, larks, and one kind of sparrow seemed to be the only bird life this grassland supported. Still, it was the wild plant life that attracted him most. Simplicity was in the eye of the beholder. Among the short-grasses George found other kinds of plants: a pigweed with minute green flowers or a peculiar kind of bee balm or a yellow flax or a hundred other inconspicuous varieties.

One day George said to Frank, "Maybe I should build my soddie into a bank, which saves sod bricks. Not to mention it heats better in the winter and cools better in the summer."

Frank laughed. "Having your soddie in a bank isn't all roses, George. Do you want a snake falling through your roof? One poor fellow had a cow fall through!"

So George surrendered that idea. Like every newcomer he started by stripping grassy sod to a depth of about four inches. The object was to make sod bricks of that thickness two feet long and one foot wide. George laid out a square fourteen feet by fourteen feet. Then following his plan he stacked sod bricks until the walls just barely exceeded his height. A man didn't want to make the soddie too spacious. It had to be heated in the winter and wood was scarce.

One day Frank drove up in his wagon. "George, I found a honey locust way over yonder yesterday. I chopped you a ridge pole. Burr oak would be better but it would cost you a dollar for a twenty-foot log."

Soon George had his ridge pole from one soddie wall to the other. From it he laid cross members of cottonwood for rafters. He could find no better wood. On the framework of the roof he laid a rough covering of sodshingles. But George was far from done. He had to frame a door and a couple of windows. Frank cautioned him to make sure the door opened inward and urged him to omit windows. George was puzzled but compromised by putting in only one window. Then he smoothed the interior walls with a shovel before plastering with the local lime. For good measure he limed the floor, too. Folks learned the hard way it was the bare dirt that swarmed with unwelcome vermin.

George looked around the clean but very barren interior of his soddie. "Now I'm going to have to fashion some furniture," he sighed, then shivered. "And I better start collecting buffalo chips for a fire this winter."

"Not on your life, George. I got you a job over on G. H. Steeley's Ranch. You're going to spend the winter in a nice warm bunkhouse . . ."

"Is winter that bad out here?"

Frank blinked in disbelief. "You mean my father didn't tell you?"

eight

"No, sir," answered George. "Your father didn't tell me."

"Last January 6th was the day the blizzard started in western Kansas. The wind was raging snow and ice on us for two days. About a hundred people died. Over twenty thousand cattle perished. Don't ever get caught away from shelter if it starts snowing out here, or even if it looks like it might snow. And get as many of the livestock under shelter as you can."

George blinked. "I know what you mean now about a nice warm bunkhouse."

The G. H. Steeleys' soddie was distinguished from other soddies George had seen by having real glass windows. Mrs. Steeley was imperious, treating George like a servant, and he had to try mightily to squelch his dislike of her. The Lord's glorious words in the Sermon on the Mount helped him:

Ye have heard that it hath been said,

132

Thou shalt love thy neighbour, and hate thine enemy. But I say unto you, Love your enemies, bless them that curse you, do good to them that hate you, and pray for them which despitefully use you, and persecute you . . .

The Lord's words were hard to obey though. George was like anyone else. His first reaction to persecution was anger that spurred him to want to fight back. But the Lord's encouragement made it possible for him to triumph over such feelings. Hating was wrong. Revenge was wrong. So he did more than just keep his feelings to himself. In his heart he tried to love Mrs. Steeley. Her absence made it easier, because soon after he started working on the ranch Mrs. Steeley went on a long trip to stay with relatives. Apparently she had no desire to be on the ranch during the winter. Mr. Steeley seemed as relieved as George when the carriage taking his Missus to the railway stop in Wakeeney disappeared on the northern horizon.

"I heard you put up a first-class soddie over on your property, George," commented Mr. Steeley.

"Frank Beeler showed me how, sir."

"Well, I surely do need your help here

putting up a soddie barn before the snow flies."

And that was the way George spent the remaining good weather of 1886. He and Mr. Steeley improved the ranch by raising a fine big sod barn. George cooked, too, constantly drawing comments of amazement from his boss and visitors. On a Monday in November, Mr. Steeley headed off to Larned for supplies that Beeler wouldn't have until the railroad reached the town. Larned was on the southern rail route of Atchison, Topeka and Santa Fe. The rail stop of Wakeeney to the north was closer but inadequate for supplies.

"Watch the weather, George," Mr. Steeley had warned before he left in the wagon at the crack of dawn.

"You too, sir," replied George.

Even though Mr. Steeley would travel the uplands between Walnut Creek and the Pawnee River, the fifty miles to Larned was precarious. There was no shelter against a blizzard out in the open plains. The trip would require Mr. Steeley to sleep one night on the prairie. No, George did not envy him. About noon the next day George was very relieved. The weather was still cool but sunny.

"Blue never looked better," said George

as he surveyed the sky. "I reckon Mr. Steeley is safe in Larned now."

Yet, hugging the northern horizon was a peculiar dark blue band. George hadn't seen that kind of blue before. But he hadn't seen mirages before he came out west, either. A fellow could be out on the prairie on a hot day and swear a lake was not far off. And if he walked to it, he never found anything but more buffalo grass. Was the dark blue band some kind of mirage? A greenish-yellow sky meant hail. He knew that much. But why was he thinking of hail in November? No, this was something different.

An hour later he scanned the sky again. "It's definitely larger," he concluded with a sinking feeling.

Every anxious moment he rounded up livestock and herded them into the barn the band grew larger. He could see the band was a cloud bank now, a deep purply-blue that piqued his artistic curiosity so much he just wanted to stop and watch the phenomenon develop. Why was the cloud bank that color? And what did it mean? Occasionally as he wrestled with a cow more recalcitrant than the others he would have doubts. Was he being the hysterical fool?

"Well, I've barned the horses and milk cows," he said at last, "and a few of the range cattle. I better haul in some fodder for them."

By the time he finished that task the wind had picked up to a steady gale. Just as he reached the soddie snow was falling. But he had seldom seen snow like this. It streaked by him nearly horizontal. And his face had never felt so cold in his life. He glanced back at the barn.

"It disappeared!"

So this was a west Kansas blizzard. George had heard some hair-raising stories by now. Ever curious he tied one end of a rope around his waist and the other end around a table leg in the soddie. Then he ventured out. Everything was white. He held his hand in front of his face. He saw nothing but white. He was blind. He backed into the soddie. Never again would he wonder how someone could not find shelter in a blizzard. His curiosity ebbed. His anxiety grew. Didn't someone recently tell him the local homesteaders had been snowbound for two months last winter? He began to wonder about supplies.

"There's not much here," he admitted after making a search. "But why else would Mr. Steeley have gone after supplies?"

The blizzard raged all night. It blew through cracks in the windows. George covered the windows with blankets. He prayed. *Yes, Lord, feed the farmer's earth moisture but don't bury the poor farmer.* He resisted the temptation to stoke up the fire too much. Buffalo chips burned very hot anyway. No, he would keep it pleasantly cool in the soddie. That way the chips would last a good long time. His wisdom was borne out the next morning. The wind had stopped. He uncovered the windows. The glass was iced over. With some effort George opened the ice-encrusted door inward. Outside the snow had accumulated a foot thick. The snow had drifted, too. The barn was a white mound.

"Thanks, friend," he said, remembering Frank Beeler's advice to have no windows in his own soddie and to make sure his door opened inward.

Mr. Steeley finally returned. The blizzard was not as bad as the one the previous January but it took its toll. The most tragic tale was of a schoolteacher in neighboring Ellis County. She and her students had been trapped that Tuesday in the schoolhouse. The children had already eaten their lunches, so they had no food. She led them in the Lord's Prayer, then wrapped

137

only in her shawl she left the schoolhouse for help. Several days later she was found under a drift, hands covering her face, frozen solid.

"Once you're in shelter, never leave it," said Mr. Steeley, shaking his head.

Why, Lord, why? wondered George. The woman was driven by duty and as brave as anyone could be. He was shaken by such tragedies. And they happened again and again. But hadn't the Lord warned that "He maketh his sun to rise on the evil and on the good, and sendeth rain on the just and on the unjust"? The Kingdom of God was not of this earth. The faithful had to muddle on, trying to live righteously. The faithful had to trust the Lord. This winter was almost as bad, said Mr. Steeley, as the previous one. It surely had winterkilled all the soft-grained wheat that had been planted.

"Too bad there's not a mill around here that grinds hard wheat," said George, remembering his conversation with Frank Beeler. "Hard wheat like Turkey Red might be able to survive this climate."

"Speaking of that," said Mr. Steeley, "I heard in Larned that the Pawnee Rock Mill has changed out its equipment so it can grind hard wheat now. Are you going

138

to try it, George?"

George mulled it over. "That's a very long haul for a wagonload of wheat. And it takes a lot of plowing to raise wheat. A fair amount of corn can be planted where I stripped the sod for my soddie. I can plow up more acreage later. It takes a lot of equipment to raise wheat and reap it and thresh it. Corn can be plucked and shucked by hand. And corn can be fed to the livestock."

The men didn't just talk agriculture the whole winter. Neighbors gathered as they had near Minneapolis to swap stories and make music. Mr. Steeley fiddled. Frank Beeler blew the cornet. Another neighbor strummed a guitar. Of course George played his accordion. They tried to harmonize on all the frontier favorites like "Buffalo Gal" and the latest rages like "Oh My Darling Clementine!" as well. Those were precious moments for the men. They knew that after spring came they would be too tired at the end of the day to do more than eat and flop in bed.

Sometimes they teased George about being a celebrity. "Says right here in the August 14, 1886, edition of the *Ness County News*," teased one friend in a pompous voice, "that 'George Carver, a young colored man and college student in the Junior class came in this week from Highland.' "

George protested, "Oh no, they got it all wrong."

His dismayed reaction was just what his tormentors wanted. Even George had to laugh. He indulged his love of plants that winter, too. He fashioned a lean-to on the south side of Mr. Steeley's soddie and somehow got dozens of kinds of plants to grow inside it that should have been asleep for the winter. The sight of his growing plants defying the season made him restless. He mulled over seeds more and more. Planting was in his dreams. Cabbage required one-half pound of seed per acre; cucumbers, two pounds; onions, six pounds; potatoes, ten bushels; muskmelons, two pounds; radishes, eight pounds; turnips, one pound . . . and to sow an acre of golden corn required about three bushels of seed. Ah, to reap twenty bushels of corn per acre! But how much seed could he afford?

"Won't spring ever come?" George would blurt out upon awakening. "I'm planting corn for sure!"

George certainly wasn't alone planting corn that spring of 1887. Ness County farmers sowed twice as many acres in corn as wheat. It was hard to argue with the numbers Frank Beeler quoted. In 1886 wheat had yielded thirteen bushels an acre in Ness County. Corn had yielded twenty-five bushels an acre.

"And wheat requires much greater effort," added George.

So in April, just about the time the railroad finally reached Beeler, George moved new-bought but sparse furniture into his soddie and impatiently waited for days that weren't too windy to scatter his hard-earned seed. He had time to bust ten acres with the plow he borrowed from Mr. Steeley, so he wasn't planting just sod corn.

He was optimistic. After all, he could grow just about anything anytime anywhere. He didn't have to wait long. He was thrilled the way the sprouts shot up in May. He watched for grasshoppers or any other bug that might threaten his green babies. How he tended his corn plants, hoeing out weeds as fast as they sprouted. Of course he also cultivated his corn, which was more or less just working the top three or four inches with a hoe. This was supposed to aerate the soil and conserve moisture. His corn would not

141

remain small long.

"It will be shoulder high by mid-June," commented Frank Beeler dreamily. "It will be all silked out and tasseled. Then your cobs will sprout their golden corn."

The corn did so well it began sending out runners which in turn shot up new stalks of corn. These renegades George chopped as readily as he would have chopped a weed. Otherwise his corn field would have choked to death on its own success. Then one morning the air was unusually warm. By mid-morning the wind blew from the south. It had been a rainy spring, so George was relieved that some of the mud patches in his field would finally dry up. The wind blew all day and ceased at nightfall. But a week later the wind was still blowing, starting about the same time every morning and ceasing at dusk. The temperature during the day was getting much hotter. Did someone say it reached ninety degrees? George would have buy himself a thermometer when he raked in the cash from his bumper crop. The wind continued to blow. George dug down into the sod. The top couple of inches were pretty dry. Some farmers were wishing they had planted sorghum, which takes heat better than anything but a cactus. Now the soil was cracking on top faster than George could cultivate it. He was very careful removing runners, so he wouldn't make the stalk's root system any smaller than necessary. Yet the wind kept blowing from morning to nightfall. The nights were now hot, too. The temperature reached one hundred degrees during the day. Where was the rain?

After more days of the hot wind George grumbled, "Lord, forget the rain. I'd settle for the shade of a cloud."

Then the corn wilted and died. And a part of George, who loved to grow things, died too. Was it the farmer part of him? No, not yet, he told himself. He was no quitter. Everyone knew farmers could have a bad year now and then. *But why, Lord?* Why such frigid winters and such hot summers? George learned the few farmers who planted winter wheat were wiped out, too. When the heads of grain were in their soft stage before hardening, somehow the hot wind baked the heads into a white doughy nothing. The winter wheat had to be burned so it wouldn't come up as volunteer. Even the sweet sorghum had died that summer. The only crop that weathered the heat was grain sorghum, the non-sweet variety in which not the canes but the seed heads were harvested. But that kind of sorghum was used for fodder. It was a poor cash crop. So George didn't torment himself by thinking if only he had sowed wheat or sowed this or sowed that he would have done all right. But he couldn't even grow a garden. It died, too. Only the homesteaders along the creeks could quench the thirst of their vegetables. This reminded George he still needed a well. That would be his fall task, he decided, when he was not working for Mr. Steeley or helping out at Bolivar Beeler's hotel.

"It's a great day for digging a well!" yodeled Frank Beeler outside George's soddie one fall morning.

George was surprised to see Frank. How did he find the time? He and the other Beelers were always trying something new on their choice creek plots. Last George heard they had given up on establishing pear trees and were full steam ahead on apple trees. They also were planting trees for timber. Walnut was a fine tree for the area but it grew slow. So they tried sugar maple, but it wilted like the corn. Now they were madly planting honey locusts, a hardy tree with good wood. In his wagon Frank had brought George the auger, an apparatus that screwed into the earth ten-foot sections of pipe that were one-half inch in diameter. There were three sections in the wagon box.

"But what if we don't find water above thirty feet?" asked George.

"Why, we just move ten or fifteen feet closer to the creek and drill again."

It seemed not to occur to Frank that they could drill holes all over George's property and never find water above thirty feet. Their first effort confirmed George's worst fears. They found no water-bearing sand. Stringers of sand were in the crumbly core but they were tight with lime. George was not so discouraged that he didn't reflect on the soil at the top of core. The upper few inches was dark gray brown loam, a mixture of silt and sand, that was matted with root fibers. This was underlain by a layer about the same thickness of similar soil but less organic and limey. Below this was a thick layer, perhaps eighteen inches or so, of the same material so limey it had white nodules. The entire upper three feet of soil crumbled apart easily in George's fingers. Where was the clay layer he had seen in other places? The soil in the upper part of the second hole was remarkably similar. George had become so interested in the soil development he didn't mind so much when the second location didn't yield water either. Why not drill on and on? Weren't they learning a lot about the uniformity of the topsoil? Finally they had to quit for the day. That evening while working at the hotel George was no longer sheltered by Frank's bubbly optimism.

"Have you run into yellow mudrock yet?" asked one "expert." "If you do, you might as well stop drilling holes, George. You'll drill all the way to China and never find water."

"Knew a fella that worked himself to death on a homestead with no water," offered another.

"It's warm down in that hole," said one man grimly. "Digging a well is a mite too close to hell for comfort." It was this man who made George's skin crawl with dread. For this man knew what he was talking about. "Helped my uncle dig wells for a while," continued the man. "Ain't no creepier feeling in the whole wide world than being in a hole four feet across, no light but a tiny piece of gray sky thirty or forty feet over your head. I'd sooner pet a rattlesnake or wrestle a bear. There's about a hundred things that can go wrong down there. The hole can cave in so fast a fella

146

knows nothing except he's going to drown in dirt. Or the rope can break and a bucket full of dirt and rock can bust a fella's head like it's a cantaloupe. That's the quickest way to die. And then there's damp gas that has no smell to it. The fella on top wonders why there's no tug on the rope so's he can lift the bucket. He finally finds out the fella at the bottom of the well is plumb dead, just went to sleep and died from damp gas. . . ."

George shuddered. Was he really going to go down in the earth like that? Well, he couldn't ask Frank to go down, could he? Only God could give him the courage he needed. That night he prayed a long time. And just before he drifted into sleep he thought he heard above the cold night wind spoken words.

The next morning he remembered those words: "perfect love casteth out fear." Was it a dream? He opened the Bible that Mariah Watkins had given him so many years ago. And there in 1 John 4 he found the words:

Whosoever shall confess that Jesus is the Son of God, God dwelleth in him, and he in God. And we have known and believed the love that God hath to us. God is love;

147

and he that dwelleth in love dwelleth in God, and God in him. Herein is our love made perfect, that we may have boldness in the day of judgment: because as he is, so are we in this world. There is no fear in love; but perfect love casteth out fear.

"Perfect love," repeated George over and over again. Trust the Lord. Sometimes men and women had to do things fear told them not to do. Digging a well was surely one of those things. . . .

Later that same day Frank yelled, "This sandy rock appears to be wet!"

George danced with joy. "Why, it's in the second section of pipe!"

"Just sixteen feet down . . . ," said Frank in an amazed voice. "And your soddie is only a couple of hundred feet away."

George began digging immediately. What a difference there was between a sixteen-foot well and a thirty-foot well. A thirty-foot well was ten times, maybe twenty times harder. George didn't even need much help. A man could easily work the hole alone for the first eight or ten feet. All he needed were a shovel and a sturdy ladder. Within two days George had reached ten feet. He looked with great fondness now at the network of grass roots in the upper three or four feet of the hole. The fibery tangle gave the hole rigidity. Frank came out to help George sink the well the last ten feet. George filled the bucket and with a rope Frank pulled it out. Of course George would have to dig down a few feet deeper than the aquifer so it would bleed water into the hole and the dirt could settle out.

"Dirt's going to start spalling off in your hole," yelled Frank from above. "The well should be walled proper. The railroad's hauling in limestone blocks from Bazine now regularly."

"Right after my first bumper crop of corn I'll wall the well," promised George. "Even before I build my stone house." George always felt optimistic when Frank was around.

But when George reached a depth of what had to be sixteen feet, the sky above seemed as small as a rifle bore. Every rock that tumbled loose from above startled him. What was that odd smell? Gas? And yet the dawning reality was worse than those fears. . . .

nine

"Where is the water?" he murmured.

George had noticed no moisture in the rocks he had been digging for sixteen feet. The same was true at a depth of seventeen feet. And eighteen feet. No water.

"Where is the water-bearing sand?" he blurted out angrily.

"Don't worry, George," bolstered Frank from above. "I'll bring the auger out and we'll keep looking."

And so they did. Again and again they augured for signs of water. With no success. George had to keep hauling water to his soddie from a spring on the Steeley Ranch. The words "knew a fella that worked himself to death on a homestead with no water" began to haunt him. Was he being foolish on this bone-dry homestead? Yet he quit working at the Steeley Ranch that winter. The plain truth was he could no longer abide Mrs. Steeley. Not that he didn't enjoy building with Mr. Steeley a hen house and a tool shed out of sod. What finally drove him off the ranch was the sale of two horses.

"Jessup and Yacob," he remembered fondly.

It had become George's job to care for the massive horses that pulled Mr. Steeley's wagon. Jessup, a lively gray, and Yacob, a white, became George's pets. How he loved to creep into their pasture and watch the two horses grazing in the distance. Finally one of the horses would sniff the air, then spot George. Both horses would bolt madly toward him to be the first to nuzzle him for a treat. How those great beasts gently clopped around him. Their eyes were huge and warm with love as he talked to them. Surely this was the way man was supposed to tend the beasts. It was like being in the Garden of Eden to George. Before the Fall.

"We've sold the horses," said Mr. Steeley one day. "The Missus thought it was best."

That raised an ire in George he hadn't felt in a long time. He couldn't help but think of his Uncle Mose back in Diamond Grove. Uncle Mose was unflappable about everything except one thing. He minded his own business about everything except one thing. If a man mistreated a horse he would hear about it from a very angry Uncle Mose. And losing Jessup and

151

Yacob made George sick inside with anger. He couldn't bear the sight of Mrs. Steeley.

All winter long George now worked at the hotel in town. But he had some time to socialize, too. He and Mr. Steeley played so well together they often provided music at weekend dances. One dollar for an evening of playing instruments and singing lively songs sat very well with George. Mr. Steeley fiddled and George played his accordion and other instruments he was mastering. The previous winter he had learned to play the guitar and this winter he was adding the organ. Some of their songs were composed locally. George had no trouble identifying with "The Little Old Sod Shanty" as he warbled:

I am looking rather seedy
   while holding down my claim,
And my victuals are not always
      served the best;
   And the mice play shyly around me
as I lay me down to sleep,
   In my little sod shanty on the claim.
Yet I rather like the novelty
   of living in this way,
Though my bill of fare is always rather
      tame,

But I'm happy as a clam in this
　　land of
Uncle Sam,
In my little sod shanty on the claim.

Was there a note of hysteria in the voices of the audience as they joined loudly in the chorus?

The hinges are of leather,
　　the windows have no glass,
While the roof it lets the howling
　　blizzard in,
　　And I hear the hungry coyote
as he sneaks up through the grass,
　　'Round my little old sod shanty on
　　the claim![1]

The amazing thing about the crop failures was that more people kept arriving in Ness County anyway. Most came by train now. George wanted to scream, "Haven't you heard? Nothing grows here!" But the population of Ness County doubled from 1886 to 1887, according to Frank Beeler. Five hundred folks lived in the township now, including colored folks. Bert Gee ran the cigar store. Clara Duncan, who had taught at Talladega College, came to homestead. And instruct. Clara was accomplished in both art and literature. So George took some painting lessons from her. She critiqued his paintings and his previous efforts that included sketches as well as paintings. He could tell she truly liked his art. That emboldened him. He had been composing poems on long winter evenings. She was the first educated person to whom he entrusted one of his poems, an epic of four-line stanzas that ran on more than forty stanzas.

One stanza read:

*The rich and poor, the great and
    small,*
By this same sickle all must fall.
*Each moment is golden and none to
    waste.*
Arouse thee then, to duty haste![2]

Clara Duncan was very blunt about his poetry, suggesting it was too labored, too enslaved to the meter and rhyme he had selected. It was better to write good prose, she added, than bad poetry. How those remarks stung him. Writing poetry was ruined for him after that. *So it is doggerel!* he fumed. But who was he angry at? Clara Duncan? He had appreciated her educated compliments about his sketching and painting. So was he now not to appreciate her educated criticism of his poetry?

"I sought her opinion," he admitted ruefully, "and by Jehoshaphat I got it."

In April of 1888, once again George impatiently waited for days that weren't too windy to scatter his precious seed in the plowed earth. Once again he was thrilled the way it shot up in May. Once again he watched zealously for grasshoppers and other bugs as he hoed weeds away from his thriving corn plants. But he was very apprehensive. The tiniest breeze sent a shudder to his soul. Then one morning in June, George realized all too well the air was unusually warm. By mid-morning the wind blew from the south. The wind blew all day and ceased at nightfall. His nightmare occurred all over again, not as windy as the previous year but just as hot and dry. Several weeks later the corn wilted and began to die. George just managed to harvest and shuck enough corn to recover some of his debt, stock some cornmeal, and buy seed for next spring. Everyone knew farmers could have a bad year now and then. *But why two years in a row, Lord?* With no water well he had no hope of keeping his garden alive either. He began to dig bounty from the earth as the Indian had taught him back in the Flint Hills. Of course he found the staple of the prairie, tipsin, which tasted like a mild turnip. Mendo, the prairie's sweet potato, was there, too. Not in abundance but — praise God — present at least in small quantities was the delicious maka ta omnicha, which he had since learned the pioneers called hog peanuts. And once he became focused, he found along Walnut Creek wild plums, elderberries, and onions. But this was not farming. This was surviving. He couldn't spend his life this way. In his poetry-writing days he'd written some lines that may have been doggerel, but at least they rang true:

O! sit not down nor idly stand;
  There's plenty to do on every hand.
If you cannot prosper in work like some,
  You've at least one talent, improve
    that one![3]

George was now twenty-five years old by his reckoning. He had not one dime of savings. All that fall and winter he worked various jobs as he deliberated. Yet at the same time he participated more than ever in local highlife. He was prominent in the literary society. He took parts in plays. He played his accordion at concerts. Again he was written up in the newspaper, this time in glowing praise that he was "intelligent and pleasant to talk to" and that his knowledge of "geology, botany and kindred sciences was remarkable," and that he was marked as "a man of more than ordinary ability." And yet the good life seemed an illusion because all fall and winter folks mortgaged their homesteads, if they had first lived on them eighteen months to prove them up. Then they boarded the Atchison, Topeka and Santa Fe with their two-dollar-an-acre money and disappeared down the track. They all insisted they would come back to their homesteads if the weather ever got favorable again. George hesitated. He could get a few hundred dollars for his 160-acre homestead. But a mortgage demanded payments with interest. Would he ever come back if he left now?

By spring of 1889 Frank Beeler was about the only one George ever saw who was still smiling. But even Frank's words were grim. "They say the population of the township is about three hundred now, George."

"Down two hundred from the five hundred last year!" ciphered George.

Bert Gee was gone. Clara Duncan was gone. But George took a deep breath and decided to give it one more try. His farming seemed like that new game that everyone was crazy about: baseball. He had two strikes on him. Three strikes and he was out. In April of 1889, for the third time George sowed precious seed in the sod. He forced himself to appreciate the way the corn shot up in May. *Thank you, Lord,* prayed George, and then added grimly, *do Your will.* George didn't pay much mind to grasshoppers or other bugs as he hoed weeds. He had never had much trouble with them. Could their microscopic brains know the crops were going to be bad? Often now he was suddenly stung with the desire to go back to school. But he had tried college and he had been humiliated. But maybe he could get a job in a greenhouse back east. He didn't have to own one to enjoy working there and studying the plants. When the hot wind came in June he was almost glad. But this time he managed to harvest some corn in the fall in spite of it. He wished he had sowed more until the crusher came. Corn prices collapsed. A bushel of corn was worth half what it was the previous summer.

"Thanks for cutting me loose, Lord," said George, finding a silver lining.

Groups of farmers were uniting to form alliances. Their rhetoric frightened George. They would repeal all debt-collection laws. They would force reduction of interest rates on loans and mortgages. They would prohibit foreclosures. Yet George could understand all these bitter desires. Hadn't he too suffered? The collapse of prices had been the last straw. He realized now all too well the variables that determined successful farming from failed farming were many, and most were beyond the farmer's control. Livestock-killing blizzards, hot, rainless summers, fluctuating commodity prices, crop-gobbling grasshoppers, and who knew what next?

Then Frank Beeler stunned him. "We're all moving out, George. All the Beelers. Beeler is almost a ghost town. I'm going down to Purcell in Indian Territory. My brother is already down there." Frank looked at George apologetically. "I wish it had worked out better. . . ."

"It will work out better," answered George. "The good Lord is surely nudging us to prosper somewhere else. . . ."

Frank Beeler and his relatives were still

159

packing up as George left. He could not tell them where he was going. He had not relinquished his homestead; he had taken a mortgage. The payments would be a constant burden, but for a while at least he would take the chance. Suppose Kansas got a few years of mild winters and rainy springs? The value of acreage would boom. He brooded over his decisions on his homestead. Could he have somehow made a go of it? Why had he deferred always to well-meaning Frank Beeler's suggestions? Why hadn't he applied his knowledge of plants more wisely?

"I know the grasses," he accused himself. "I know that only buffalo grass and grama grow well where the soil is very dry. Very similar in form to those two is another grass: wheat. Therefore shouldn't wheat grow well where it is dry? On the other hand corn is a grass in form most like bluestem grass that grows best where the soil is more moist. So corn needs moisture. . . ." George faltered. "On the other hand, sorghum also is a grass very similar to corn and bluestem, yet sorghum resists dryness. Was it sorghum's more waxy leaf that resisted drought? Oh, blazes!"

He gave up in frustration. Besides his analogy breaking down, he could not articulate his certainty of the similarities and differences among the grasses. He knew there had to be a way but he lacked the education. And what about the differences in root systems? What about the way each kind of grain formed and when it formed? So he brooded. What would have happened had he planted Turkey Red wheat as he had originally planned? Would he have had a crop in spite of the scorching south wind? Could he have wagoned his hard-grained wheat all the way to the mill near Larned and still made money? Or could he have survived by planting grain sorghum, the only crop he knew of that could really resist the hot summers they had suffered from 1887 to 1889? And who would have bought it? Oh, the plight of the farmer.

"And yes, the plight of George Carver, too," he groaned. "What is to become of me? I've failed at everything I've done but my schooling and yet I can't go on to college. . . ."

In Wakeeney George heard the railroad was hiring colored men to cook for the rail-laying crews. The main lines had been laid, but now the railroads were building spur lines to small towns off the main line. Impulsively George hired on. Why not go farther west? First he had to convince the railroad officials he knew how to cook. But the proof was in the pudding and soon after George cooked for the officials he was climbing into a boxcar with a rail-laying crew. The boxcar lurched ahead, and with other cars loaded with rails and ties, it clackety-clacked west behind a cinder-spewing locomotive. Between the burgs of Grinnel and Carlyle the foreman of the crew slid the door open a foot or two to point out where the three hundred renegade Cheyenne Indians led by Little Wolf and Dull Knife had crossed the Saline River.

"When did that happen?" asked one suspicious worker.

"Don't rightly know," replied the foreman.

George knew. It had been 1878.

Soon the railroad people massed together into a flurry of men and activity, building a spur line. First, carloads of crushed rock were laid out for roadbed material. Heavy wooden ties saturated with creosote were laid next. Then a railroad car with a huge crane lowered steel rails. Metal plates were arranged between the tie and rail. Dozens of men hammered in spikes so that the rails were firmly attached to the ties. The railroad cars carrying all the men, supplies, and equipment rolled right over the freshly laid track. Everywhere George explored the countryside in his spare time.

Near Denver, George left the railroad. He headed south. He scrubbed clothes, ironed shirts, chopped wood, dug gardens, trenched, and in every menial way possible survived. Finally, George was all the way down into New Mexico. True desert was another new experience. Plants clumped together in a sea of what seemed hard, barren soil. But there he saw what seemed to him the most beautiful plant he had ever seen. Rising from an impenetrable mass of green spikes was a stalk covered with white cup-shaped blossoms. The plant was a blooming yucca. How he wanted to paint it.

"Always the plants seem to thrill me the most," he admitted, but did he most want to paint the plant or did he most want to study it and understand it?

How he began to crave the lush greenery back east now. So he headed back. He stopped off in Minneapolis to see Aunt Lucy and Uncle Ben. But he didn't make the rounds. His rejection at Highland College was too embarrassing to some there.

163

George just didn't embarrass friends and acquaintances if he could help it. He could still write his friends. But he had to find out from Uncle Ben how the local farmers fared during the terrible hot years that plagued him at Beeler. Uncle Ben confirmed that 1887 had been a disaster in Ottawa County too, both corn and winter wheat yielding less than ten bushels an acre. The following year corn yielded a puny ten bushels an acre, but winter wheat skied to twenty-one bushels per acre. The next year, 1889, winter wheat continued to improve, yielding twenty-four bushels per acre. To make wheat even more attractive, it was bringing sixty cents per bushel.

"Some folks say a man would be a fool to plant corn in the drier parts of Kansas," avowed Uncle Ben. "Especially with all the droughts and poor corn prices."

George couldn't argue with that. But the vagaries of the weather and the economy had wilted his ambition to be a farmer. Farming fascinated him, but oh the heartbreak! George said his farewells and drifted east. Leaving the state of Kansas — with scant intention of returning — was not lost on him. He had arrived in the state January 1877 a fearless untested boy of thirteen or so, and he was leaving nearly thirteen years later a man of many accomplishments — but also of many failures. His onetime dream of helping his race now seemed a million miles away.

By late fall he was in Winterset, Iowa, cooking at the St. Nicholas Hotel. Winterset was only a one-day walk from the much larger town of Des Moines, but the truth was that ever since Fort Scott, George had avoided larger towns, with the exception of his few weeks in Kansas City in 1885.

"Winterset has just enough folks living in it to be civilized yet friendly," he said with satisfaction.

Never had a name for a town seemed more apt. He was hunkering down for the winter, sure that his situation as a cook was not permanent. One of his favorite Scriptures was Proverbs 3:6. "In all thy ways acknowledge him, and he shall direct thy paths." It never failed him. So he attended the Methodist Church. And George wasn't bashful about singing. Hymns were sung for the Lord's glory. It was not right to hold back. After the service the choir director, Mrs. Milholland, invited him to her home. After George arrived he complimented the very large and well-furnished home.

"My husband is a doctor," she explained

165

simply. "George, you have a wonderful tenor voice. Will you sing for me at the piano?"

So George sang. Once Mrs. Milholland joined him. He stopped abruptly. "Ma'am, your soprano is so controlled, so modulated, I am embarrassed to sing with you. I must sound like a squealing hog."

"Nonsense. I want you to sing in our choir. If you think you need a couple of lessons I will give them to you."

George said, "I will most happily accept your offer to sing in the choir, ma'am. But I'm afraid I must decline your lessons. I can't afford them."

"But they would be free."

"No, ma'am," he said firmly. He was still very touchy about getting something for nothing.

While he was there Mrs. Milholland showed him her painting. How he envied her. Her stool and easel had their own area by a large, well-lit window. Even now her palette and brushes rested on a small table beside the easel which supported a canvas painted with flowers. George realized her painting was the equivalent of his own poetic doggerel. Her rendering of the flowers — no doubt the result of much love and enthusiasm —

was a failure nevertheless.

"Well, what do you think?" she asked expectantly.

"Ma'am, perhaps you should lighten up this portion here," suggested George tactfully.

"Show me," she said skeptically.

So George picked up her palette. Within seconds he deftly applied and blended white to highlight stalks and leaves and petals. Depth entered the painting. But more needed to be done. Totally absorbed, he continued. He darkened some areas, enlarged others, corrected some botanical details. Suddenly he was jolted by the brashness of what he was doing.

"Oh excuse me, ma'am!" he blurted. "For a moment I thought I was doing my own painting."

"But I'm amazed, George! My painting has come alive."

Before George left that day they had struck a bargain. Mrs. Milholland would give George voice lessons. He would give her painting lessons. Shortly after this arrangement began George lost his job at the hotel. Once again he had to start taking in laundry and ironing. But it did give him freer hours. Almost every evening he was at the Milhollands'. Mrs. Milholland not only taught him to sing, but tried to train him to speak in a deeper voice. No one would have been happier than George to lose an embarrassingly falsetto voice. That was the reason he was always selected for ladies' parts in plays. But deepening his voice made him feel like a phony. Mrs. Milholland constantly prodded him, assuring him practice would make him less self-conscious. Soon his well-modulated tenor speaking voice would be second nature. But try as he might George could not force his voice lower for more than a few minutes. There were far too many other things to think about.

"Good heavens, ma'am," he blurted out uncharacteristically, "I can't go around the rest of my life worrying about the pitch of my voice."

So Mrs. Milholland failed in that regard. Otherwise, George's relationship with the Milhollands was a blessing. Dr. Milholland and the two children were as cordial to George as Mrs. Milholland. George even celebrated Christmas in their home. He confided in them about his crocheting and his love for plants, as well as his desire to paint. When spring of 1890 came he had little desire to move on.

One day in the presence of her husband Mrs. Milholland shocked George by saying, "George, this will never do."

"What do you mean, ma'am?" asked George, sensing trouble.

ten

George, you have far too much ability not to be in college," replied Mrs. Milholland.

"I tried that once, ma'am," answered George uncomfortably.

He had never confided in them about Highland College. George was not a complainer, but even if he had been a complainer he probably couldn't have talked about Highland College. It was just too painful. And why bother the Milhollands with the racial prejudice set in the hearts of some people? The Milhollands had treated him with complete respect. Even love.

Dr. Milholland spoke. "Well, I don't know what happened before, George. And it doesn't matter. It's my diagnosis that you're definite college material and I intend to see that you go."

"I pay my own way, sir," objected George.

"Oh, we know you well enough now, George, not to offer you anything you don't earn," said the doctor. "But it

170

wouldn't be wrong for me to open a door for you, would it?"

"No, sir, I would not object at all," said George, thinking that might well be a door not even the good, well-intentioned doctor could open.

But George was wrong. The rapidity of developments that summer of 1890 dazzled him. Once again he was scheduled to matriculate at a college in September. This time the college was Simpson College, a Methodist institution in the town of Indianola, twenty-five miles east of Winterset. But George was very apprehensive. The last time he reported to matriculate was one of the most painful experiences of his life. Even Dr. Milholland's assurances couldn't comfort him. Only his prayers at night comforted him. It was then he calmed and remembered the kindnesses of folks like the Carvers, Helen Eaker, Doctor McHenry, the Beelers, Mr. Steeley and others toward him, not out of pity but out of love. He must not let the hate of a few poison his heart. God would judge those who tried to hurt him. Yes, the Ultimate, the Almighty would someday judge him and all others, too.

"But this time I'll not count my chickens

before they're hatched," he reminded himself as he walked to Indianola.

No, he was not going to stroll into Indianola like he had strolled into Highland, dreamy eyes noting oak and hickory, curious hands finding a peculiar flowered herb. No, this time he was not going to anticipate morning hikes, nor plant sampling, nor delicious mysteries of nature. *Do Your will, God,* he prayed. And he steeled himself as he reported to the registrar. *Oh please God,* he prayed, *don't let this person blurt out "why, you're colored!"*

"George W. Carver, hmmm . . . ," said the registrar, pulling out a file. "I see our president Reverend Holmes has deemed your high school certificate from Minneapolis insufficient. . . ."

"Insufficient, sir?" George felt sick. Simpson College was more subtle than Highland College. He was being prepared softly for rejection. "Insufficient, sir?" he repeated numbly, his voice more high-pitched than ever.

"You lack the necessary mathematics." The registrar smiled. "That only means that you are being admitted on probation. You must take a mathematics course with your other courses. So the president has registered you for mathematics and several studies of English, including grammar and composition."

"I would like to register for botany, too, sir."

"I'm sorry but we don't offer botany. Our courses are very heavily liberal arts."

"Then I would like to register for art, sir."

"Art? Why, only . . . only our young ladies generally take art courses."

"But I would like to take art, sir."

"Perhaps you could discuss it with the instructor, Miss Budd. If she will admit you then I suppose it will be all right."

George found that peculiar but he scurried off to the art department. He entered the main art room where he was told Miss Budd could be found. He was aghast with envy. Easels and chairs were everywhere in the huge room. Sketches and paintings adorned the walls. Best of all, the room was brightly lit by a skylight. George felt like he was in a greenhouse. It was paradise.

"Yes?" said a soft voice behind him.

George turned. The speaker was a young lady in the attire of the day: full, ankle-to-neck dress with long sleeves. But her muslin dress was distinguished by a cherry red ribbon and its rich plum color.

"Ma'am, I'm looking for Miss Budd."

"I'm Miss Etta Budd. What can I do for you?" She was frowning.

"Ma'am, I'm George Carver. The registrar said I must get your permission to take an art course."

"But why ever would you want to take an art course?"

George blinked disbelief at the question. "Ma'am, I love art."

"And what have you done in art?"

George explained what he had sketched, what he had painted under the direction of Clara Duncan, and that he would love to paint plants. Miss Budd raised an eyebrow at that last remark.

"Plants? Indeed. Sit down there at that easel, sir, and with that charcoal sketch me a rose." George quickly rendered not only a blossom with five heart-shaped petals but the leaves and the stem. Miss Budd studied it intently. "Sir, I must tell you it doesn't look like any rose I've ever seen. The petals aren't right. Nor are the leaves. And the stem looks prickly rather than thorny."

"Ma'am, it's a prairie rose. In the month of June. Oh, let me correct this." He hurriedly altered the center of the blossom. "This part should be flatter."

"Well, I don't recall ever seeing a prairie rose, but you seem certain that sketch represents one. All right, I'll let you attend my class for two weeks, George Carver."

George was elated. But remaining in art class was only one of his tasks. He had no place to live. Men students had no facilities at the college. And he had to find work. But he was a past master at surviving. Over the next few weeks he had passed muster in the art class, found a room near the college, and set up his laundry business. His confidence and his optimism grew greater every day. Much of his success was due to Miss Budd and Miss Budd's friend, Mrs. Arthur Liston. But even with their help George had only boxes for furniture and cooked on a stove he rescued from a dump. Simpson College had three hundred students, of which one — George — was colored. Except for an occasional snub George felt as comfortable as he had in Minneapolis and Ness County. And he fastidiously protected his friendships, even acquaintanceships. If he saw one of the other art students with someone he didn't know he avoided them, so the art student would not suffer possible embarrassment by having to speak to him. Was he right in behaving this way? Was he a coward? A traitor to his race? All he knew was that life for him was very complicated.

"You must come to our house for Christmas, George," said Mrs. Liston.

The Listons owned a book store and they became George's unofficial host family in Indianola, as the Milhollands had been in Winterset. George maintained his friendships in other towns with lively letters. With the Milhollands he felt so comfortable he offered blunt advice like "if you are thinking of investing much in orchids I think you will regret it,"[1] or made requests like "send me a bulb of that oxalis we got from Miss Siders last fall."[2] But once in a while he became too full of himself:

(Miss Budd) . . . paid me two very pleasing compliments today. She said that in all her contacts with people she never met anyone like (me) and that I was going to excel in flowers. She further told me she had never seen me get them stiff and ungainly, and she is not going to let me copy a thing, but make my own designs and paint from nature . . .[3]

Then he had to temper his self-praise by adding "this subject is getting very monot-

176

onous."[4] But he unabashedly admitted to the Milhollands, "I realize that God has a great work for me to do."[5] Never had his optimism been so great. Still, usually he signed his letters "Your humble servant in God."[6] He begged God's forgiveness, too, because the truth was that for a long time after Highlands College he had doubted the promise of Philippians 4:13, "I can do all things through Christ which strengtheneth me." But God had preserved George's desire to go to college in his heart — like a seed in the long dry years of the desert — awaiting just the right moment.

George found out with certainty that some of the other students liked him. One day he returned to his room to find his furniture suite of boxes pushed aside and in its place a bed, table, and chairs. And the donors knew George well because the gifts were anonymous. Otherwise George would surely have returned them. George suspected they also conspired to give him tickets to music concerts and plays by claiming their friend got sick.

"It will just go to waste," the benefactor would assert. "Could you use it, George?"

Besides Mrs. Liston, George's other great friend in Indianola was Miss Budd.

Their relationship was teacher and pupil, and Etta Budd was as intensely interested in his future work as George was. He learned why she was so reluctant to let him into her classes when at last she voiced her concerns. Why was he committing himself to art? Didn't he know the opportunities in art were few and far between for colored people? Would he help others of his race even if he did succeed in indulging his art? What of his love of plants? Couldn't he apply that love to better the lives of colored people much more effectively?

"Ma'am, I'm torn between that and exercising my rights as one individual who happens to be colored. My Uncle Mose said a fella has a right to be different. Uncle Mose is different in his odd ways — not going to church and such — and folks respect him anyway. Now you're telling me I'm not an individual but one of my race."

"No, you can remain an individual, but you can also advance the cause of your race. Just as I'm an individual and hopefully I can also advance the cause of women. Right now teaching art is opening up for women and I intend to flood America with women art teachers! Call it a 'stepping stone' for women if you like."

"I see, ma'am." George shook his head.

"But Simpson College doesn't teach anything about plants. . . ."

Miss Budd smiled. "But the Iowa State College up in Ames does."

"I see, ma'am," said George.

George trudged back to his room. Did he dare dream of studying plants? Hadn't he dreamed often of working in a greenhouse? Wouldn't he learn all the names he had craved to know for as long as he could remember? Wouldn't he learn how to graft and cross-fertilize and such? Was Miss Budd right? Did he love plants as much as he loved art? Perhaps in the back of his mind he had thought the science of plants was more closed to him than art. It was hard to remember. So many different doors had closed on him so many times. George's mind swam with possibilities. His heart was warmed by painting. But the thought of plants gave him joy, too. More? Less? He couldn't tell. At Simpson College George was less than fifty miles south of Ames where Iowa State was located. But he knew that he was separated from Ames by much more than fifty miles of road. There was a great chasm there that separated the aspiring artist in George from the aspiring scientist in George. Agronomist? Botanist? George didn't know what he wanted to be. He lacked the knowledge and the vocabulary to articulate his desires. He had recognized that lack since he was child. And he had real doubts about his ability to succeed at Ames. Whereas he was beginning to feel competent in art he felt only doubt about agricultural science. Hadn't he failed as a farmer? And at Ames wouldn't he have to struggle through the adversities of mathematics, chemistry, physics, and who knew what other subjects before he ever saw a rose?

"There is no shame in wanting to be an artist," he argued with himself. "But what about my race? Will my art help my race? And how much am I obligated to help my colored brothers and sisters?"

For years he had vacillated between a desire to enjoy himself as an individual and his desire to help his race. Back in Minneapolis he had soared with ambition to help his race, even planning a book. Then Highland College brought him crashing to earth! Would rejuvenating this ambition bring him crashing to earth again? What did the Bible say about obligation to one's own kind? The sage in Ecclesiastes hinted at it: "Woe to him that is alone when he falleth." Paul, in 2 Timothy 4:16, seemed indirect too: "No man stood with me, but all men forsook me: I pray God that it may not be laid to their charge." John 15:13 was more relevant: "Greater love hath no man than this, that a man lay down his life for his friends." Any hope George had of brushing all this aside as ambiguous was lost when he read the Lord's parable in Matthew 25:

And the King shall . . . say unto them . . . Depart from me, ye cursed, into everlasting fire, prepared for the devil and his angels: For I was an hungred, and ye gave me no meat: I was thirsty, and ye gave me no drink: I was a stranger, and ye took me not in: naked, and ye clothed me not: sick, and in prison, and ye visited me not.

Then shall they also answer him, saying, Lord, when saw we thee an hungered, or athirst, or a stranger, or naked, or sick, or in prison, and did not minister unto thee?

Then shall he answer them, saying, Verily I say unto you, Inasmuch as ye did it not to one of the least of these, ye did it not to me.

Who were "least" than colored folks? He went back to Miss Budd. "How do I

181

find out about Iowa State?"

"I'll have information sent to you."

The more George learned of Iowa State the more excited he became. Iowa State was the nation's very first land grant college. When President Lincoln signed the land grant College Act in 1862 that established one agricultural college in every state, Iowa was the first state to be ready. Iowa State was on the cutting edge in agriculture, creating an experimental creamery in 1879 and an experimental agricultural station in 1888. It was closely tied to the United States Department of Agriculture, and the best part, George learned, was that no admission was denied because of creed, sex, or race. The applicant needed only to prove he was a scholar. Miss Budd assured George his case would be heard.

"My father, Dr. J. L. Budd, is the professor of horticulture there," she finally admitted.

"Horticulture!" yelped George. "The science of growing fruits and flowers! Why, Miss Budd, you amaze me."

But George was still apprehensive. He loved Simpson College. The students accepted him and offered him friendship. Suppose he was unqualified for Iowa State? Suppose he went to Iowa State and was turned away. The memory of Highland College still stung him. Or was it the memory of how proud he was? Every time he thought of his triumphant tour of Olathe, Paola, Diamond Grove, and Neosho he felt so foolish. So why should he keep blaming Highland College? And wasn't he engaging in self-pity? And what did the Bible say of self-pity?

"There probably is no proverb about self-pity," he told himself, "because it is so totally unacceptable among God's people."

In May 1891 George appeared at Iowa State. The nine-hundred-acre campus was on the heights in rolling timbered hills. The buildings were diverse, with bricks of blond and red, round Italian cupolas, and English watch towers. Classes were in session already. The school year of two semesters ran from Washington's Birthday in February until Thanksgiving in November. At North Hall, Professor Budd warmly welcomed George. He gave him his class schedule and told him where he could buy his books. George attracted little notice until some male students realized that a yardman or a janitor would not be carrying books.

"We don't want niggers here!" shouted one.

The next blow came when George learned he did not have enough money to pay for a dormitory room. He returned to Professor Budd at North Hall. What was he to do? Dr. Pammel, Professor of Botany, solved the problem by installing George in an empty office. Realizing the full scope of George's financial needs Professor Pammel also arranged for George to do janitorial work in North Hall. But George's problems were far from over. The next blow came when he was informed by the manager of the dining hall he could not eat with white students in the dining hall but must eat in the basement. Had he abandoned Simpson College for this? Very depressed, he wrote letters to his good friends Mrs. Milholland and Mrs. Liston about his first day at Iowa State. In his misery he held nothing back.

"Mrs. Liston!" exclaimed George one morning several days later. He had rarely seen her dressed so splendidly. In fact she was highly conspicuous. She not only carried a brightly-colored parasol, but wore an enormous floppy hat with a ostrich-feather plume. Her dress seemed fit for a royal ball.

"Can you take me on a tour of the campus?" she asked.

All day George escorted Mrs. Liston around the campus. She went into the dining hall and had a pleasant discussion with the dining hall manager. Did he know of George's many accomplishments? Iowa State was lucky indeed to claim him as a student. Before they left, the dining hall manager insisted George have his meals with the other students in the main hall. She went with George to North Hall and spoke to students, rattling off George's accomplishments there as well. Her effort was transparent. Would it work? George noticed the day after her departure his treatment by many had improved to at least accepting him as a student. Once the other students included him, he soon won them over with his humility and his astonishing experiences.

"George here survived a west Kansas blizzard," remarked one student in the dining hall. "What was it like, George?"

George knew how to spin a yarn. "When I woke up one morning and saw the mice were wearing earmuffs and snowshoes I knew I was in trouble. . . ."

Given a chance George could almost always win friends. But his stay at Iowa State improved immeasurably when he joined an evening prayer group. Not only

185

did he communicate his deepest concerns to God and other students but he met Professor James Wilson in the prayer group. The deeply devout Wilson had just come to Iowa State himself to head their experimental station. Professor Wilson was fifty-five, and an esteemed scientist. George was in awe of Wilson. So were all the other students.

The prayer group grew so large Wilson had to split it into more groups by "graduating" some of the students. One of the graduates was George. But the scheme failed. George and all the others who were graduated drifted back to Wilson's prayer group. Wilson seemed a patriarch right out of the Bible. He was righteous, but not other-worldly. Wilson not only knew about plants but he could communicate his knowledge. He not only could manage his own affairs but he could manage the affairs of others. He was a teacher, a leader, an organizer — and righteous. George felt honored when Professor Wilson pulled him aside one day.

"George, we must never allow what happened to you in your first days here at Iowa State to happen to other students. We need to get some of the older students to meet incoming freshmen and shepherd

186

them to the registrar, their rooms, the dining hall, and all around campus. And of course they will be invited to prayer groups. And you are the one student mature enough to organize that effort."

George was thrilled, so much so that he had to use Paul's inspired words in 2 Corinthians 3:5–6 to quell his self-importance:

Not that we are sufficient of ourselves to think any thing as of ourselves; but our sufficiency is of God; Who also hath made us able ministers of the new testament; not of the letter, but of the spirit: for the letter killeth, but the spirit giveth life.

George not only headed the welcoming of new students, but became a delegate in the Young Men's Christian Association. So the grim first days at Iowa State faded in George's memory. Once again he was accepted for who he was. He was most certainly regarded as different — but it was not because of the color of his skin. . . .

eleven

George was different. He was *the* botany student on campus.

Virtually every morning he was seen going on nature hikes. It was a rare day George did not greet the rising sun. Always he walked about campus with a metal container hanging from a shoulder strap. In this homemade contraption he would save any interesting botanical specimen he might spot — afternoon or evening, as well as morning.

His conversation with a colleague over a tree must have seemed Greek to others. "Yes, sir," reasoned George, "I agree the leaves are odd — pinnately compound with lanceolate leaflets. But the margins are serrated, and note the density of the panicles. Surely this sumac is the stag-horn variety and not the shining variety."

Oh, how many years had he longed to speak such a precise language? He gloried in botany and horticulture. He seemed to do a thousand other things as well. George performed in plays, usually as a siren-voiced lady. He helped the football and track teams as an equipment manager and masseur. He organized an Agricultural Society, debated in the Eclectic Society, practiced his painting and sketching in the Art Club, and learned a foreign tongue in the German Club. Although military training was required at a land grant college he advanced from the ranks to drill the cadets as their Captain, the highest rank attainable by a cadet! Military training also mandated a change in George's posture. Years of toil and solitude had hunched his lean frame forward. But by locking a stick behind his back with his arms during his long hikes he was soon walking tall again.

In the meantime over his four years at Iowa State George toiled in English, elocution, law, history, algebra, geometry, trigonometry, surveying, geology, zoology, bacteriology, entomology, and heredity. He squeezed in fun courses like drawing and oration. He took many hours and levels of an essential subject: chemistry. And his knowledge of agriculture bloomed in courses on grasses, livestock, dairy farming, plant pathology, nutrition, and farm drainage. And of course he took all permutations of his beloved botany and horticul-

189

ture. In his favorite subjects rarely did he receive a grade other than an A. In the other courses it was rare for him not to get at least a B.

After his first months at Iowa State he had rarely encountered racial prejudice. Any overt antagonism toward him by a new student was met with immediate and very rude reprisals from other students. Discrimination was not tolerated, officially or unofficially. Times were getting better for colored people, but nationally they were far from just. George knew he was at the forefront of change. His behavior influenced white people, eroded their prejudices. He knew that. He had witnessed it. And he protected the feelings of others. Never would he embarrass anyone. He was truly "wise as a serpent." For years he had been the epitome of tact. Only occasionally at Iowa State could he completely forget all the taboos that concerned him. Once, while digging plants with a knife, he responded to an underclassman's repeated impudence by chasing him.

"If you cut my jugular vein I'll never forgive you," screamed the laughing underclassman.

In December 1892, Iowan artists were exhibiting their work in Cedar Rapids. One purpose was to select work to represent Iowa at the World's Fair in Chicago the next year. Many on campus knew George had some remarkable paintings of flowers. "Why shouldn't he go to Cedar Rapids?" they asked. He told them such a trip for him was an extravagance. He could barely make ends meet and he was still paying on his homestead. Besides that, he couldn't spare the time. "You owe it to the honor of the school," they insisted shrewdly, and by chicanery they lured stubborn George to a clothier. Even Dr. Pammel was there. Protesting, squawking George was outfitted in a new suit, gloves, and hat. Days later his kidnappers forced him onto a train to Cedar Rapids with four of his canvases, including his huge canvas titled *Yucca and Cactus*. In Cedar Rapids all four paintings were selected as part of Iowa's representatives to the World's Fair in Chicago.

"We knew you could do it!" cried the masterminds of his abduction after he returned.

"Praise God you don't think your money was wasted then," George told them.

Because George loved his *Yucca and Cactus* most of all and it alone was more than a handful he decided he would take only it to Chicago. His abductors arranged

191

to have him photographed with his huge painting. How he loved these students and professors. Such a congenial group had seemed beyond his reach. Now he bathed in the radiance of their respect and kindness. The students were wonderful — and had there ever been such a faculty? Men like Wilson, Pammel, Budd, Stanton, Wallace, Osborn, Byer, Marston, Curtis, Knapp, Stalker, Bennett, and Patrick. All so kindhearted and wise. And not all men, either. George was also taught by Misses Roberts, Doolittle, and Stacey, all stalwarts, too.

It was no surprise George finally relinquished his homestead. No longer did he entertain any thought of ever going back, good times there or not. At Iowa State he became aware of literature on the Kansas prairie that had already been written before he homesteaded. Martin Allen's extensive experiments near Hays showed the water-impoverished high plains had a dismal future for corn. Louis Agassiz, famous professor of geology at Harvard, had taken one look at the high plains and insisted only wheat could be grown there by dry-land farming. T. C. Henry's experiments with wheat at Abilene seemed to confirm that. Well, time would tell, concluded George, if wheat or any other crop would profit a farmer on the high plains of Kansas. What if he had possessed all this knowledge before he had homesteaded? Well, almost certainly he would not have made the mistakes he did. In fact, he probably would not have tried it at all. With no mill nearby to grind hard wheat, the only grain crop with a good chance of success, the area was foredoomed.

He mentioned this to his professors, too. "Why aren't common folks informed of these farming facts?"

His interests became more and more professional. In 1893 he published his first scientific paper in the Transactions of the Iowa Horticulture Society. "Grafting the Cacti" expressed not only his burgeoning knowledge but his bedrock philosophy: every plant has a myriad of uses; it is for man to discover those uses. He pointed out for the cacti that "doubtless their widest range of usefulness as yet locked up within them" await "the kindly hand of man."[1]

One year later at the annual meeting of the Iowa Horticulture Society he presented a paper titled "Best Bulbs for the Amateur." That same year of 1894 in his senior thesis, "Plants Modified By Man," he described state-of-the-art work on plum trees, geraniums, and amaryllis in grafting and cross-breeding to improve fruit, vigor, and appearance. Much of his own work had been on the amaryllis, simply because a bulbed plant could be transported so easily. At the mature age of thirty, he expressed in his thesis the desire he had felt since a boy. Why shouldn't the horticulturist blend and

> *perfect the color of his flowers, producing not only harmony, but a glorious symphony of nature's daintiest tints and shades, with just as much certainty as the artist mixes his pigments upon the palette . . . ?[2]*

Yes, at long last George had acquired the skills to fulfill his childhood dream of mixing flowers!

George received his Bachelor of Science degree November 14, 1894. He had no cheap degree, amassing 145 hours of mostly A's in very difficult coursework. Graduation moved him into a world he had never known before. All the menial work he had performed to survive was behind him. The faculty persuaded him to pursue a Master's Degree. He would earn a salary as a graduate assistant, with his main duty that of tending the greenhouse. There he could indulge his urges to create new plants. His graduate work was under Professor Pammel, who had helped him from his first day at Iowa State. Pammel's specialty was fungi, a field called mycology. George threw himself into his new work. Not only did he sketch and paint specimens of fungi, but in the outdoors he sought new specimens for Professor Pammel. He not only found numerous mushrooms and other fungi in the Ames environs that were rare, but several that were unknown to science!

Professor Pammel didn't mince words. "George, you are without a doubt the best collector I have ever known."

Often George took professors' children with him while collecting. One of the more precocious was Henry Wallace's six-year-old who carried the same name. Little Henry was fired with enthusiasm by George. "Yes, yes! I see it!" little Henry would scream. In the next two years George contributed fifteen hundred specimens to Professor Pammel's collections as well as collaborating on three scientific articles on fungi. As his graduation once again approached, this time in the fall of 1896 for a Master's Degree, George had a decision to make. Should he take the offer of a position with Alcorn A & M, a college for colored folks in Mississippi, or should he remain at Iowa State? For in no uncertain terms Iowa State wanted him to remain. James Wilson declared that in cross-fertilization and the propagation of plants George was the ablest student at the college. But Wilson's praise went far beyond that. George was stunned to learn Wilson had written Alcorn A & M:

Except for the respect I owe the professors, I would say he is fully abreast of them and exceeds in special lines in which he has a taste . . . I would not hesitate to have him teach our classes here . . . We have nobody to take his place and I would never part with a student with so much regret as George Carver . . . It will be difficult, in fact impossible, to fill his place.[3]

George respected James Wilson above all other people. Wilson was the only one in the world who could cry, "Look at your shoes, George! You can't face the undergraduates that way. Now go buy a new pair right away!" And off George would obediently trudge with the dollars Wilson had thrust upon him. But George learned Wilson himself might leave Iowa State. Faculty were hearing a rumor Wilson would become the Secretary of Agriculture under William McKinley if McKinley won the presidential election in 1896. George was delighted James Wilson might assume such a powerful position. Imagine what such a wise man could do for agriculture in America. But the thought of Wilson departing Iowa State depressed George as much as suddenly seeing some great oak tree gone.

George echoed Wilson's words but now they were about Wilson himself. "It will be difficult, in fact impossible, to fill his place," muttered George.

Meanwhile in April 1896 George received another offer from a college for colored people. Tuskegee Institute was a school in Alabama, administered by the best known and most widely respected colored man in America: Booker T. Washington. Washington realized Tuskegee needed a first-rate agricultural program and through his inquiries he had discovered to his amazement that well-informed academics said the most ingenious horticulturist in America was a young colored man! That man was Iowa State's George Carver.

Only seven or eight years older than George, Booker T. Washington was also born a slave, but on a plantation in Virginia. Following the Civil War, he too burned with desire to be educated. By 1879 he not only had a college degree but he was teaching in college. Then he was tapped to found Tuskegee Institute in 1881. Under his leadership Tuskegee was becoming an important college for colored people. In the process he became an orator much in demand. His nationally-known Atlanta speech in 1895 urged moderation and accommodation. He implored colored people to learn vocations and climb the ladder to economic freedom. Whites of course applauded his attitude. Colored people were divided. Some like W. E. B. Dubois demanded much quicker justice. But George Carver, the epitome of patience and tact, couldn't have agreed more with Washington's moderation. In fact, it seemed Washington used the very metaphor "step by step on the golden ladder" that George had imaged back in Minneapolis.

Still George was cautious. Just what kind of agricultural program did Tuskegee have in place? On April 12, 1896, he wrote Booker T. Washington:

. . . it has always been the one ideal of my life to be of the greatest good to the greatest number of "my people" possible and to this end I have been preparing myself for these many years; feeling as I do that this line of education is the key to unlock the golden door of freedom for our people . . .[4]

Yet George still had not accepted the offer.

Washington wrote him that the John F. Slater Fund would pump money into a new agricultural program at Tuskegee, even erecting a new building. He also added that if George did not accept the offer he would have to fill the position with a white man. This went against Washington's grain because he thought the symbolism of an all-colored faculty was very important to the students. George agreed. On May 6, 1896, he wrote to accept the offer from Tuskegee. He also assured Washington that he "said amen" to all the accommodation Washington proposed in his speeches, which was "the correct solution to the race problem."[5]

George left Ames before the formal graduation ceremony for his Masters Degree. But he was far from empty-handed. Besides his clothing, books, canvases, and many cases of specimens he had a magnificent new microscope, one more generous gift from his friends at Iowa State. The love that he felt from the faculty and students at Iowa State had finally completely broken down his reluctance to get "something for nothing."

"To refuse the microscope would be to refuse their love," he admitted.

And to refuse the microscope would be to deny Tuskegee Institute a microscope it might badly need. On October 8, 1896, he arrived by rail in Tuskegee, a small town forty miles east of Alabama's capital, Montgomery, and one hundred fifty miles north of the Gulf of Mexico. It was in a province of once rich longleaf pine forests, now riddled with cotton fields. When George entered the two-thousand-acre campus of Tuskegee Institute he saw shops busy with smithies, wagon makers, tailors, carpenters, shoemakers, clothiers, brick makers, and broom makers. Tuskegee Institute was, after all, a vocational school. Although all-brick Alabama Hall stood four stories high, much of the Institute was crude, reflecting a shortage of funds. Not only were the roads dirt, but the yards as well. Nighttime lighting was provided by oil lamps. Hordes of birds and flies scavenging in a ditch near the dining hall betrayed the lack of a sewage system. But George saw only possibilities. He would make the bare soil of the campus green with flora. Tuskegee Institute would become an oasis for wretched colored folks. George would teach them how to return to their farms of sorrow as masters. If they were too backward to come to Tuskegee he would go to them.

"Welcome, Mr. Carver," greeted Booker T. Washington.

Washington was tall, wiry, square-jawed, and very articulate. But his greeting was strained; he was distracted. George began to feel uncomfortable with him, as if he were keeping Washington from a thousand tasks. So he excused himself. The all-colored faculty was cool to George, which puzzled him. It wasn't his imagination, either, because they were extremely cordial to each other — too playful, in George's opinion. Perhaps they resented his generous annual salary of one thousand dollars plus room and board. Perhaps it was his exalted title of Director in Scientific Agriculture and Dairy Science. Perhaps it was the fact he would direct five instructors who taught horticulture, stock raising, farm management, truck gardening, and dairy farming. Perhaps it was the arrival of many boxes containing his art work and botanical collections. Perhaps it was the fact he was the only one on the faculty with a degree from a "white" college. But George had endured cool receptions before.

In the first days he informed another faculty member, "Last night I ate in a colored restaurant nearby."

"You are a Yankee, aren't you?" The faculty member laughed bitterly. "You didn't have to say it was colored. If you ate at a restaurant in Alabama it was colored."

In the same bitter tone George was taught etiquette for a colored man among whites in Alabama. On the sidewalk he should always pass a white person on the gutter side, preferably stepping off into the street. His eyes should always be averted from theirs. He should take off his hat when speaking to any white person. He should never attempt to shake hands with a white person. He should never address any colored person as Mister, Missus, or Miss; those appellations of respect were reserved for whites and demanded by them. He must end every sentence to white men with "sir," to white women with "ma'am." He should never be out walking after dark. Nor should he try to register to vote. Violating any of these requirements would drive one of these white men into rednecked rage, even to the point of knifing or shooting the offender. And white men here were never convicted by all-white juries of violence against a colored person. Besides the injuries or death the colored person might suffer, reminded one of the faculty, the offender might sour the white townsfolk on the Institute and Dr. Washington would not like that.

"I see," said George, too numb to add comment.

In spite of the cultural perversities, George was thrilled to hear of the progress at the Institute. Not the least of its development was the planned construction of the new Slater-Armstrong Memorial Agricultural Building. In the meantime he must get students interested in agriculture. The truth was that colored people in Alabama considered farming about equivalent to slavery. After the Civil War the southern whites learned how to unofficially enslave colored people through sharecropping. The white landowner provided land and equipment. The colored farmer provided labor. They shared the crop. But the colored farmers lived barely above the level of slaves. Colored people were trying to escape farming. George found he had only thirteen students enrolled in agriculture!

"This may be more difficult than I thought," he admitted to himself.

George began to familiarize himself with the land. The coastal plain here was somewhat unique in that it was originally underlain by dark clay soils derived from limestone. For that reason it was called the Black Belt although many folks now thought it was named that because within it "black" people outnumbered whites. The cash crop was cotton. Much of the topsoil had been lost, George was told, by the reckless clearing of the pine forests. Cotton itself had depleted the richness of the topsoil. Many fields were virtual badlands, barren and gullied. More stark yet were the multitude of gray unpainted shanties of the sharecroppers. Wandering about on the bare dirt of the tiny yards were toddlers, pigs, and chickens. Occasionally on a porch George saw an adult, usually a woman barefoot and ragged, too pregnant to work in the fields. This was October. All the others, young and old, were in the fields picking cotton.

"This one-crop system must end,"

declared George.

To this end he drew up a plan for an experimental station and presented it to Washington. If he had any doubts about Washington's abilities they ended when the very plan he drafted became Alabama law just weeks later in February 1897. A branch Agricultural Station was to be established at Tuskegee under the direction of its Director and Consulting Chemist, Mr. Carver! George wanted to reciprocate. He was pleased to learn that very month James Wilson had indeed become the new Secretary of Agriculture under newly elected President McKinley. George wrote James Wilson. Would his old professor be able to find the time to come to the dedication of the new agriculture building?

In the meantime George worked on the internal design of the new two-story building as well as recruiting students. George's experience in plays and debates had made him a relaxed and fearless speaker. His education and experience only reinforced his confidence. *If these students just give me two minutes of their time,* he bolstered himself, *I can convince them that I can teach them to farm.* George was too shrewd to flaunt his education and use big words. Besides, his supreme mentor was the Lord, who spoke in simple parables.

"We belong to a very powerful race," George would declare to listening students firmly, then boom, "if we organize! There is a not a single animal or a single plant in nature that is not organized. Think about it! The very act of creation was God organizing the heavens and earth from void and darkness. He proceeded very deliberately and never finished any one thing of the millions of things he created without saying finally 'that is very good.' And so are we one of those millions of things and just like the all the others 'very good.' Yes, God made us 'very good'!" George held up his hand, fingers spread. "Like this my hand cannot perform one task out of the many it is designed to perform. I cannot grasp. I cannot lift. I cannot even make a gesture until my fingers organize and work together! That is why I am here. I am going to help you, made by God 'very good,' to organize you so you can work!"[6] He closed by shaking his fist.

And how well his pep talk worked!

Students flocked to him. At the rate he was attracting students he expected to have over one hundred for the next school year.

206

He was so persuasive he even had three girl students. He had never been so busy. He rarely socialized. There was no time left in his day to do that. Not only did he have students to recruit, a curriculum to organize, floor plans to design, classes to teach, instructors to administer, field and garden crops to tend, orchards to harvest, beehives to keep, and poultry and dairy cows to care for, but he had his own botanical interests to satisfy!

That latter pursuit was how the white people of the town of Tuskegee discovered him. . . .

twelve

"Say there, boy," bawled a well-dressed woman standing in the backyard of her estate, "just what are you doing over there in those piney woods?"

George Carver removed his hat and called back from the woods, "Collecting plants, ma'am." He had been putting specimens in the ever-present container that hung from his shoulder.

"Whatever do you know about plants, boy?" she asked.

"I've studied them in school, ma'am."

"Indeed," she said skeptically. "Well, come and have a look at my azalea bushes. The leaves are turning brown."

George immediately saw that her soil did not drain well enough for azaleas. After the wealthy woman had her azalea bushes raised slightly and the soil enriched with organic matter, her azaleas thrived. And George found himself being called upon by townsfolk with increasing frequency. His expertise drew invitations to give talks on plants. Booker Washington was delighted that George was a good diplomat for the school. But the drawback was that George had even less time to win friends among the faculty. Some seemed only to resent his popularity with white folks. And his accomplishments in art and botany, which he guilelessly discussed, they regarded as bragging — or perhaps even lying. With effort George could have smoothed things over. He had always been able to make friendships. But he had no time to make the effort. What little time he had left during the day he gave to the students. And the students returned his commitment, which many in the faculty resented even more.

George was not shy about asking for what he considered his just due, either. "I must have more room for my valuable specimens," he complained to Booker Washington. "I came here with no other motive than to benefit our people."

"Some of the faculty make only half what you make," explained Washington impatiently.

The coolness of his fellow faculty members made George treasure the memories of his years in Iowa even more. Often he wrote Professor Pammel and others on the Iowa State faculty. He recounted to them no story more often than the one where Pammel and others virtually kidnapped him, clothed him, and sent him off with his paintings on the train to Cedar Rapids. This memory of their great love for him always brought tears to his eyes. He was thrilled when the professors there now asked him to write a short autobiography of his early years, implying he was someone future historians would want to know more about.

Of course he wrote often to the Listons and especially to the Milhollands:

> *You, of course, will never know how much you . . . (helped) a poor colored boy who was drifting here and there as a ship without a rudder. You helped to start me aright and what the Lord has in his kindness and wisdom permitted me to accomplish is due in a very great measure to your real genuine Christian spirits. How I wish the world was full of such people, what a different world it would be . . .*[1]

He wanted only to express gratitude to old friends, but he couldn't keep from hinting at the pain he felt from prejudice of all kinds.

When the 1897 fall term started George learned his request to James Wilson had been accepted. Wilson would come in December to help dedicate the new agricultural building. Complaints from the faculty about George's presumptions were silenced. George let white townsfolk think Booker Washington had attracted the secretary. And the townsfolk were impressed. The Secretary of Agriculture coming to Tuskegee? What a great honor for the community. Booker Washington was only too willing to let the townsfolk share the spotlight. But James Wilson drew others as well. The governor was coming. So were local politicians, clergy, merchants, alumni, newspaper reporters, so many that the crowd swelled beyond the one thousand students and faculty to five thousand in all.

"George," asked James Wilson privately, "are you finding time for your plants?"

"Yes, sir. Especially now that we have an experimental station."

And so he did find time somehow. That spring the students had plowed by ox about twenty acres. This acreage was divided into many experimental plots. A business had donated several hundred pounds of fertilizer. But George knew few colored farmers could afford commercial fertilizer. He had to seek alternatives. One way was to add organic matter. This was possible perhaps for a garden, but not on a scale large enough for a field crop. No, George had to find another way to restore vigor to the cotton-depleted soil.

"Legumes," he concluded.

Legumes were a family of plants with an enormous number of individual species. Some were woody plants, some were herbaceous, some even aquatic. Their most noticeable characteristic was a seedpod that split open easily to reveal one or more seeds. But their trait that attracted George was the ability of their roots to take nitrogen from the air and turn it into nitrate — a fertilizer!

"But which of the over ten thousand species will do best for a farmer here?" he wondered.

Legumes included trees like acacias and mimosas. Also mesquite, brazilwood, carob, honey locust, Judas, logwood, and tamarind trees. But no matter how well-intentioned, trees were not an option for a money crop. Other legumes seemed to come closer to the mark. Why not clover and alfalfa? His inquiries had discovered no clover in the entire county. Or why not broom, and lupine? Or peas, beans, and soybeans? And peanuts?

"Peanuts? I'd as soon plant parsley," commented one farmer.

But George's students were soon planting various legumes. And he certainly could not neglect cotton. If some were determined to raise cotton then he would just have to try to develop better varieties. Nor could he neglect other crops. One constituted his most amazing achievement thus far at Tuskegee. In the second of his Experimental Station bulletins, which he intended to publish four times a year, for free if possible, he described how productivity of sweet potatoes could increase with proper enrichment and cultivation of the soil. The average yield per acre in the Black Belt for sweet potatoes was 40 bushels. George's plot yielded 266 bushels per acre!

Dan Smith, an accountant from New York who audited Tuskegee's accounts, didn't disguise his awe of George. "How do you do all these things, Doctor?"

"I'm not a 'Doctor,' " explained George. "I have only a Master's degree."

"Baloney. Your work merits you a doctor's degree."

Dan Smith kept on calling him Doctor Carver. The unearned title caught on.

213

George's reputation spread as "Doctor" Carver. After a while he gave up protesting its use. That was a distraction; after all, he was being called upon as a problem solver. Could he please analyze this soil? Would he mind testing this water? Could he recommend a cure for an ailing horse? In the fall of 1897 George had started meeting the third Tuesday of every month with local farmers. There they could air their problems and pick his brains. The meetings were called the "Farmer's Institute."

Other colored people were shocked when they saw a white farmer hold out a wilted leaf, almost contritely, and ask, "What do y'all reckon ails this here plant, Doctor?"

George also became interested in local clays, so glaringly exposed in gullies. Yet they rainbowed beauty, too, from snow-white to yellow-ochre to the richest sienna and Indian reds. Perhaps the Lord had exposed them so men could put them to good use. George wondered if he could mix his own pigments. One red clay had glints of blue and he isolated that rarest of pigments. George took out a patent for his blue pigment but it wasn't specific enough to prevent a national paint company from jumping on his discovery. This George did not see as a tragedy but as a great benefit to Alabama. But he wasn't through with clays. He was certain white clay could be used as whitewash. Why should the wood-frame houses of poor farmers be unprotected? And for that matter, why was their furniture unfinished? He was sure some of the clays could be used as stains. And he had seen some very pure kaolin. Why couldn't they make ceramics? Even fine china? Oh, if only he had a potter's wheel. Perhaps he could make his own.

"Is it any wonder I have no time for socializing?" he commented.

In 1898 the Institute was honored by a visit from President McKinley and his entire cabinet, including, of course, Secretary of Agriculture James Wilson. George was sure his friendship with James Wilson had made it possible. Then in 1900 Booker Washington's autobiography, *Up From Slavery*, was published. Of course George read it, surprised to see almost no mention of Washington's staff at Tuskegee except Mr. Logan and Washington's brother John. The slight seemed unnecessary but George could accept that. Mr. Logan had served at Tuskegee a very long time. And who wouldn't mention their only brother? But Washington's proud account of the Institute's new Slater-Armstrong Agricultural Building angered George. Not only was George not mentioned as head of the agricultural studies but Washington took sole credit for securing James Wilson's visit in 1897 to dedicate the building. George scarcely needed to read the account of President McKinley's visit to Tuskegee to know it was all to the credit of Booker T. Washington. The book killed any delusion George had that Washington was in some way going to promote his career.

But George faced much greater problems. His trips to other towns to give talks were not only time-consuming but often dangerous. One such trip was to the tiny town of Ramer, south of Tuskegee. When he arrived, a white woman photographer was already there putting together material on a book about colored schools. Her guide was the colored teacher at the local school, a man named Mr. Henry. After being introduced to the two fact-gatherers George left them and returned to the house where he was a guest.

Later, the front door was hammered. "Let me in!" screamed someone. The door was opened. It was the woman photographer. "They're chasing Mr. Henry!" she yelled. "Shooting at him!"

"Who is?" asked George.

"Vigilantes."

The memory of Fort Scott turned George's heart to ice. None of his old western adventures gripped him with fear the way the sick madness of lynching did. He had no idea if Mr. Henry escaped or not. Oh, how George wanted to hide until the vigilantes tired of their sick prejudices. But he must help the woman photographer. George prayed for strength. Later that night he drove the photographer to the next railway station, then returned to Ramer. But the incident was not over. The next day one of the vigilantes, armed with a shotgun, mistook George for Mr. Henry. Luckily he did not shoot before George proved who he was. Even then, the vigilante questioned him sharply about who might have spirited the photographer out of town. Upon learning Mr. Henry had escaped, George left town. The colored school was not spared. The vigilantes broke the windows, turned over desks, and ripped the books apart.

"So education for poor colored folks takes one more step backward." George felt crushed.

Administrative duties at Tuskegee were

crushing him, too. Booker Washington turned out to be a great stickler. The Institute must be perfect in every aspect, he insisted. He criticized George for a poorly placed sign, a piece of machinery left outside — and a dozen other things. Once he assigned George to investigate and identify the culprit who had been driving a buggy recklessly though the campus. But the thing that became a major problem for George was the poultry farm. The inventory was always less than it was supposed to be. This loss infuriated Washington. But how was George to stop the theft of chickens? Nevertheless it became such an aggravation to Washington that it opened the door in 1902 to the ambitions of a newcomer to George's department.

G. R. Bridgeforth tried openly and unrelentingly to get George removed as head of the agricultural department. George tried just as hard to get Bridgeforth removed from his department. But George had even fewer friends on the faculty since he had been allocated two rooms in the boys' dormitory. Other unmarried male teachers were two to a room! And the newcomer did demonstrate management skills. Plus he concealed a vicious streak from all but George. He provoked George with notes:

"You must do business like a man," and "Let us be men and face the truth!"[2] When George complained to Booker Washington, the latter praised the newcomer's boldness and initiative.

"I'll write James Wilson," George grumbled finally. "Maybe he knows an institution where I can do some research."

Wilson advised George to stay put. Where else could he help his own people more than at Tuskegee Institute? George should try to get his teaching load reduced, suggested Wilson, so he could have more time to do his research. There was no one in the world that George admired more than James Wilson. So George tried to push the strife out of his mind and continue his myriad of activities. Yet his anger seemed to permeate a leaflet he produced for the Institute in 1902:

The average southern farm has but little more to offer than about one-third of a cotton crop, selling at two and three cents per pound less than it cost to produce it, together with the proverbial mule, implements more or less primitive, and frequently a vast territory of barren and furrowed hillsides and wasted valleys . . . The southern farmers have been too slow to admit that the old one-crop and primitive implements are quite out of harmony with the new, and up-to-date methods and machinery. Indeed many are not aware that such conditions exist . . .[3]

Ironically George now had many friends among the faculty. One of the faculty wives began promoting her sister as a match for forty-year-old George. George was confused. Should he at long last marry? He sought advice in the Bible. It shamed him to presume he might be like the saints. But what had been their attitude to marriage? Peter was married. Apparently Paul was not. Either way was acceptable. George never felt less decisive. He prayed for guidance. He wrote Mrs. Liston for advice. She encouraged him. Then the decisive answer came. George's love interest delivered it with hot words. Any wife of George's would be a widow no matter how long George lived, she declared angrily. He traveled more and more, she pointed out. When he wasn't traveling he was in the classroom or in the fields. Even when he was with her, she said hotly, his mind was rarely available. Thus the courtship ended.

George, stung at first, was in honest moments relieved. "The truth is I have no time to be a good husband."

In 1906, with donations from Morris Jesup, George had the students build a Jesup Agricultural Wagon, a wagon loaded with farm equipment, soil samples, placards for illustrating lectures, and anything else useful for conveying information to farmers. All summer long George toured with the wagon, stopping in dusty town squares but just as often stopping on a lonely clay road abutting a cotton field. In this way he provided farm information to about two thousand listeners a month. "Only two thousand?" wondered some skeptics. "Why not just put out a bulletin?"

"Because many of these farmers can not be reached that way," George insisted. "They can't read."

As if George didn't have enough to do the students talked him into leading a Bible class every Sunday evening before chapel. "But don't the boys and girls get their fifteen-minute devotionals just before their classes begin and again in the evening before study hour?" objected some. George couldn't argue with those facts but neither was he going to argue with the thirst some students had for the Lord's Word. Never had he been so aware of Paul's words in 2 Timothy 2:15: "Study to show thyself

221

approved unto God, a workman that needeth not to be ashamed, rightly dividing the word of truth." Because George was able to reconcile the Bible with scientific fact to his own satisfaction, he felt obligated to help others with that supposed problem, too. Soon the study class swelled to eighty and it had to be moved into a lecture hall where George expounded on maps and charts and geological specimens.

"You must learn to read the great and loving God out of all forms of the existence He has created. Converse with God. Oh, what a joy will come to you. Oh yes, God is so lavish in the display of his handiwork!"[4] Sometimes George was so overcome by the sudden memory of the glory of a sunset or even a rock or a weed he had to pause until he could calm himself.

By 1908 Bridgeforth had convinced Booker Washington to split the agricultural department into two divisions. Bridgeforth would head one called Agricultural Industries. George would head Agricultural Instruction and Experiment Station. This freed George to do more rewarding work, but he was hurt anyway. It was almost like losing his fiancée. His pride screamed in pain, but his heart knew it was a blessing.

222

Now he could really do some botanical research.

But George soon realized the division had been a ploy by his rival. Bridgeforth informed Booker Washington that the split was not going to work. The divisions had to be reunited into one department again, he insisted, but he should be the head, not George. George could remain to teach a few courses and supervise the Experiment Station but he would report to him. To George's dismay Washington agreed! George now was to have only the Experiment Station and teach a few courses. But once the changes took place, his teaching load was not reduced. To placate him, Washington promised George he could equip a first-rate laboratory for himself. But this became a source of aggravation for George because his requests for equipment were constantly delayed. After eighteen months he wrote Booker Washington an angry letter that expressed far more complaints than the aborted laboratory:

> *For sixteen years I have worked without a single dollar's increase in salary . . . I have sat quietly and seen others go way above me in my own department . . . (bulletins are) practically impossible to get*

223

published . . . not a single piece of (bacteriological) apparatus purchased . . .[5]

He went on to detail all the things that had been promised him by Booker Washington and remained unfulfilled: reduced teaching load, laboratory equipment, a museum, a chemist, a stenographer. He mentioned the utter contempt a few of the faculty — probably poisoned by Bridgeforth — now expressed to his face about the laboratory. If George ever received his laboratory, which was doubtful, they sneered, and left Tuskegee later they would surely let his laboratory fall into ruin because it was of so little use to anyone!

About this time Washington's book *My Larger Education* was published. It lauded George, citing how impressed certain foreigners were with his accomplishments. One was Sir Harry Johnston, who had visited Tuskegee. Washington quoted Sir Harry's book *The Negro in the New World*, in which he declared that George's science was as sound as that of any professor at Oxford or Cambridge, the two great universities of England. In a few moments of conversation, Johnston wrote, any botanist would realize George was their equal.

Still, George was too soured by his

demotion to appreciate Washington's recognition of his accomplishments. And when George walked around the campus now he could scarcely appreciate his own efforts in transforming it from bare dirt to green expanses of Bermuda grass. George's disposition wasn't improved by receiving an obituary. Moses Carver had died at nearly one hundred years of age in Galena, Kansas. Aunt Sue had died years before. Both George's "white parents" had been very good to him. Aunt Sue had not only taught him skills that had enabled him to survive hard times, but she had encouraged him to love plants. Uncle Mose had taught him a fellow could be different and still stand tall. How he loved their memories and lamented their passing.

In spite of failing to get his teaching load lessened at the Institute, George expanded his own teaching effort outside the Institute. He could not neglect the teaching of colored children. They must know the world itself was a classroom. He even wrote a manual for teachers. He emphasized this kind of teaching must begin with the youngest tots. They must learn to identify bugs and rocks and such things. They must grow their own little gardens, care for

225

their tools, prepare their soil, fertilize, select seeds, plant, cultivate, and market. Using the outdoors like this helped the child relate to all other subjects taught.

"You just wait till that child finds a bug eating his cabbage plant," said George with a chuckle. "He will write a composition about it with a vengeance! And if he can sell you a carrot or two he will happily do the math to tell how much you owe him."

Ignoring his difficulties at Tuskegee, George built a local reputation as the "plant wizard." Of course he experimented, too. He couldn't maintain the pace of four informational bulletins a year. With the difficulty in actually getting a bulletin into print he barely got out two a year. But they were significant: "How to Build Up Worn Out Soils," "Feeding Acorns," and "Cowpeas" were but a handful. Often an intended bulletin would never reach fruition because of the many possibilities his initial investigations indicated had to be pursued. But never did George dream that in 1903 the subject of one of his initial investigations would yield amazing potentials that exceeded those for any plant he had ever known. . . .

thirteen

"The peanut," said an amazed George.

The story of George Carver's fascination with the lowly peanut really began with an even lowlier insect less than one-quarter-inch long. George had long known there was a beetle that fed on cotton plants, not only in its long-snouted adult form, but in its larval form, too. The munching maggots hatched from eggs the beetle laid inside the bolls of cotton. From egg to adult was a mere three weeks, so the cotton crop had to survive generation after generation of greedy beetles. The devastating beetle came to be know as the boll weevil. Back in the 1860s it had prevented the growth of cotton in Mexico. By 1893 it had crossed the Rio Grande to infest Texas cotton.

"Folks keep saying the spread will stop for this reason or that reason," said an alarmed George in 1910 to his colleagues, "but it just keeps coming closer and closer. Every year since 1892 it has spread about fifty to one hundred miles farther north and east."

"So?" asked one dense listener.

"So farmers around here better plant their cotton real early and pick real early. Maybe they can beat the weevil that way. Meanwhile I'll work on a better variety of cotton, one that ripens earlier yet. Maybe I'll think of a crop they can plant instead of cotton. . . ."

"Instead of cotton?" blurted one listener. "Why, that's what folks grow around here!"

George was fond of the sweet potato as an alternative crop and he had already experimented with it extensively. But he had another option, too: the peanut. He had already experimented with it, as well, because it was a legume that restored nitrogen to the exhausted Alabama soil. George became enthralled with the peanut. It grew like a weed, rain or no rain. Most farmers permitted a vine or two to grow around their shanty just so their children could eat the raw seeds they called goobers. George scarcely needed to improve the prolific plant. No, George had to demonstrate there was a market for peanuts and for that he had to find uses for peanuts. One could not grow them instead of cotton just because some entrepreneurs had begun to roast them and sell them at a few baseball games.

"They are African, too," he said with appreciation, "brought to this soil by slave ships."

But that was not the complete truth. *Arachis hypogaea*, of the Fabaceae family, had been cultivated since ancient times by Indians in South America. Peanuts were exported to Africa, then came to North America with African slaves. Plants grew about thirty inches tall and bushed out to a width nearly twice that. The plant seemed the most bizarre on the planet. After the flower was fertilized it turned downward on its stalk and buried what was now the "peg" into the soil! There the tough, fibrous pod with its one to three bean-like seeds developed.

George imagined a myth for his affair with the peanut. It was one he would tell to his audiences. It began by him asking God why he made the universe.

"Ask for something more in proportion to that little mind of yours," replied God.

"Explain to me why you made man," asked George.

"Your little mind still wants to know far too much. For your modest proportions I will grant you the mystery of the peanut. Take it inside your laboratory and separate it into water, fats, oils, gums, resins, sugars, starches, and amino acids. Then recombine these under my three laws of compatibility, temperature, and pressure. Then you will know why I made the peanut."[1]

And that was what George — a very gifted chemist — did.

First he concentrated on its food derivatives. By 1913 he issued a bulletin on the many ways to prepare peanuts as food. But he was making discoveries so fast the bulletin had to be revised again and again. Ten recipes, twenty recipes, thirty, forty, fifty, where would it stop? Still, if the peanut was to be accepted as a viable crop by farmers it would be none too soon because the boll weevil plague was getting closer and closer to Alabama. The year 1913 also brought in a new American president, Woodrow Wilson. It marked the end of James Wilson's sixteen-year reign as Secretary of Agriculture. The new president and his new Secretary of Agriculture were indifferent to Tuskegee. Booker Washington had to work even harder to raise money for the school.

In 1914 the boll weevil problem dimmed in comparison to a war raging in Europe. Funds for the Institute dried up because of the war and Wilson's unresponsive administration in Washington, D.C. Booker Washington was gone from the Institute all the time trying to raise money. America was supposed to stay out of the war but who could tell? But didn't the Lord say "Take therefore no thought for the morrow. . . . Sufficient unto the day is the evil thereof"? As if to prove the Lord right once again George had an incident that summer in one of the newfangled horseless carriages. A "truck," the driver called it. George was riding in back with several others. The driver lost control and the truck flipped. George was pinned underneath the wreckage. Sickened by the salty smell of blood, it seemed he might die from being crushed, but the wreckage was finally levered off its victims. George was bleeding and bruised. It seemed miraculous that at his age he had no broken bones. No one died in the accident, but one man was so severely injured he would probably never walk again.

If 1914 was bad, 1915 was worse, a year of reckoning for Alabama and Tuskegee Institute. The cotton crop was devastated by boll weevils. All predictions paled before the facts. George promoted peanuts now at a furious pace, though few farmers did more than listen. Then in November Booker Washington, only fifty-nine but exhausted from ceaseless fund-raising and worry, suffered a heart attack on a trip north. He left the New York hospital under a death sentence. He lasted only long enough to return to his home at the Institute.

His death devastated George. For years Booker Washington had seemed the thorn in his side. Now George had to reflect. Had George been fair to him? Perhaps not. The man had been under enormous financial pressures. No one complained louder than George when in 1914 Washington asked all his faculty to go one month without pay. Not only had Washington been worn down by squabbling at the Institute and by raising money outside the Institute, but he had been vilified by militant colored people as a weak-kneed "Uncle Tom." Now George felt very guilty about his persistent complaints. A good man was dead. Washington was far too young to be dead. He had died for the cause. Washington never knew how much George loved him and his noble cause. How could he know when George had complained so much? And George had been so sour he had scarcely acknowledged

Washington's glowing praise of him in *My Larger Education*. Had George been fair to him? Definitely not. Guilt hung on George like iron chains. He prayed to God for forgiveness. At the funeral in Tuskegee former President Teddy Roosevelt gave a wonderful eulogy for Booker Washington. Roosevelt was about Washington's age and in ill health himself. That made George feel even more rotten about himself and his complaints. He trudged off to be by himself after the funeral.

He had a visitor the next day. "Mr. Carver?"

It was Teddy Roosevelt!

"Yes, sir," said George numbly.

Roosevelt stuck out his famous bulldog chin. "Listen to me good now," he said. "There's no more important work than what you're doing right here!"[2]

Roosevelt said his farewell and left. George was stunned. Had his face revealed his troubled spirit at the funeral? That was doubtful. George had long ago developed a poker face in response to prejudices. Had his old friend James Wilson, Secretary of Agriculture under President Roosevelt as well as McKinley and Taft, told Roosevelt of his troubled protégé? Yes, that must have been it. James Wilson, himself about eighty, must have said to Roosevelt, "Look in on young George Carver because he's been troubled for some time now." Imagine!

"Let's get on with our work," said George determinedly. " 'The sluggard will not plow by reason of the cold' or by a thousand other reasons — yet all are better reasons than my self-pity!"

George began to function again, and yet with that came his old complaints. In spite of his misgivings over his quarrels with Washington, he was soon complaining about his teaching load to the new president of the Institute, Dr. Robert Moton! But how could he not complain? There was so much else besides teaching he had to do. And once again he was refused. Moton said there was no way to get a replacement for George's teaching load on such short notice, not that George could be replaced.

"Stick it out through the spring term of 1916," asked Moton.

But the next year was the same. A replacement for George could not be found. Again George had to assume his heavy teaching load. Still, George's research yielded his best bulletin yet on peanuts. And farmers were listening. Many acres of peanuts were being planted, especially i Coffee County, about fifty miles to the south. At last George would have some results from real farmers to tout. Often in his jaunts into the country a farmer would reveal a peanut recipe. Each revelation was like finding a jewel. One time a man innocently said his mother made coffee from peanuts.

"Coffee?" exclaimed George. "I've never been able to make mock coffee!"

The man explained how his mother roasted the peanuts, then cooled them before roasting them a second time. George tried that method. It did yield good mock coffee. George had so many recipes now that he had a cooking class prepare a lunch of soup, salad, mock chicken, creamed vegetable, bread, candy, cookies, ice cream, and coffee — all made from peanuts!

That fall of 1916 his world seemed to explode with good fortune. First he was notified that he was elected to the advisory board of the National Agricultural Society. That honor was quickly followed by an invitation to be a fellow in the prestigious Royal Society of Great Britain. How had that happened? George could only suspect he had been sponsored by Sir Harry

Johnston, the visiting naturalist who had lauded George in his book *The Negro in the New World*. Johnston had declared George's science equal to that of any professor at Oxford or Cambridge, hadn't he?

The faculty of the Institute was astonished. Moton was delighted. With Booker Washington dead, most worried the Institute was cut loose and adrift from national support. Now it seemed they might once again have a national figure, if the newspapers picked up on it. Moton made sure many newspapers were informed by telephone of George's accomplishments. Local newspapers willingly reported this colored scientist's achievements. Then finally a few national newspapers and magazines began to report his story. Under the circumstances George was hardly surprised when at long last Moton reduced his teaching load. George was politely asked to teach one course for the fall term of 1917, just to maintain contact with the students. George agreed, although he lived in the boys' dormitory and still conducted his very popular Bible class.

His tension with the rest of the faculty eased. One reason was that he was very popular with their children. He had long taught them at the Children's House, their school at the Institute. He took them on nature hikes. He even guided them through his laboratory — if they agreed to keep their hands behind their backs! Thomas Campbell's house was one faculty home where George was always welcome for supper on holidays. George came early. Hidden in his pockets were peanuts, which the Campbell children would root for. Then George romped and roughhoused with them.

After supper he would sigh, "Well, I'm going to leave this 'boiler factory' now."[3]

His ascendancy and popularity were too much for G. R. Bridgeforth. George's old nemesis left Tuskegee. It seemed now half the boy babies born at Tuskegee were named George Carver This and George Carver That. George's fortunes kept improving. In 1919 Moton increased his salary. George hadn't even asked. Although salary had been a sore point with him in the past, the truth was he often forgot to cash his paycheck. Far more important to George in 1919 was Coffee County. What a showcase it was now for peanuts. Because its farmers had switched to peanuts it was one of the few counties in the state prospering. At the county seat of Enterprise local businessmen had constructed a shelling plant. They had also put up a monument to the boll weevil! Yes, it was this little pest — and George Washington Carver — that pushed them into their new prosperity.

In May 1920 the peanut farmers of the southern states met in Atlanta to organize the United Peanut Association of America. Peanut farmers still had many problems. There was no large processing plant in the south. The market was weak. And they had foreign competition for what little market existed. Peanuts imported from China were nearly as cheap as their own. The tariff on imported peanuts was so small it was insignificant. One of the goals of the new peanut association was to convince the federal government that they needed protection. One member, noting George Carver's extensive work on uses of the peanut, suggested George be invited to give them a talk.

"A colored man?" asked one farmer aghast.

Nevertheless, George was invited to their meeting in Montgomery, Alabama, on September 14, 1920. The thrust of this meeting of the United Peanut Association of America was to gather information to take to Congress in an effort to get a tariff on imported peanuts. U. S. Senator Underwood and U. S. Representative Heflin were at the meeting, too. The room was tense when George rose to speak to a sea of skeptical white faces. Some had heard of George's peanut milk and were disbelieving. Many were prejudiced against colored people and in no way considered them equal. George proceeded to talk to the stony-faced men about the uses he had found for peanuts, illustrating with samples of peanut products he had lugged there in several cases. The audience remained silent. Had George failed in his mission?

The next to speak was Congressman H. B. Steagall, who said meekly, "Following the speaker who has just addressed this meeting I certainly do feel out of place and greatly embarrassed, and especially so if I attempted to talk about peanuts in any way to instruct anybody." But after more remarks Steagall asserted strongly, "When the time comes that this question must be thrashed out before the American Congress I propose to see that Professor Carver is there in order that he may instruct them . . ."[4]

George was startled as the audience erupted into applause. The association went on to unanimously resolve to help George in any way they were able, including securing patents for his products. George's head was swimming with the suddenness of his triumph. And the men kept their word. In January 1921 the United Peanut Association of America paid his way to Washington, D. C., to speak before the Ways and Means Committee of the House of Representatives. The subject was Tariff Revision. Joseph Fordney of Michigan chaired the meeting. When George's turn to speak finally came the members of the committee were not in a good mood. It was not only a Saturday meeting but it was late in the afternoon. They had reviewed tariffs on various farm crops all day long. The need to recess was written on all the weary faces. And who cared about peanuts anyway? Would the committee postpone George's presentation?

"Or worse yet, cancel it . . . ," he murmured.

fourteen

Chairman Fordney said irritably, "All right, Mr. Carver, we will give you ten minutes."

"Mr. Chairman . . . ," George nodded politely.

But his mind raced. Only ten minutes! It usually took George several minutes just to unpack his samples. So he talked pleasantly and tried to keep his composure as he unpacked his many vials full of stains and other peanut products. "I am going to just touch a few high places here and there because in ten minutes you will tell me to stop. . . ." He continued to unpack his samples and talk.

"If you have anything to drink, don't put it under the table," quipped the Chairman.

"The drinks will come later if my ten minutes are extended."

The members laughed. Several began talking.

"Let us have order! This man knows a great deal about this business." The angry reprimand came from John Garner of

Texas. He was intensely watching George's samples.

George held up chocolate-covered peanuts. "As I passed through Greensboro, South Carolina, I noticed in one of the stores that this was displayed. I am very sorry that you can not taste this, so I will taste it for you." The members laughed again. George could see they had stopped looking at their pocket watches. Some were even sitting forward in expectation. Of what use was that amber liquid in that vial? What was that small cake? George continued, "The peanut and the sweet potato are two of the greatest products that God has ever given us. They can be made into a perfectly balanced ration. If all the other vegetable foodstuffs were destroyed a perfectly balanced ration with all the nutriments could be made from the sweet potato and the peanut. From the sweet potato we get starches and carbohydrates, and from the peanut we get all the muscle-building properties."

A Congressman from Connecticut made a racial slur about having watermelon for dessert and George crushed him with the reminder they all had done very well without dessert during the Great War. It was the last tasteless joke that Congressman would make. George doggedly explained his samples. He held up a very popular product, salted peanuts. Then a peanut bar with a sweet potato syrup matrix. He showed hay made from the plant after the peanuts were harvested. Peanut hay was as nutritious for livestock as alfalfa. He demonstrated that hulls could be used to burnish metal. He explained that peanut hearts, which had to be removed before processing for human food, were ideal for birdfeed. He displayed ground peanut meal for making flour. He showed several kinds of breakfast food from peanuts. He revealed the skins off the peanut could be used to make thirty dyes shaded from orange-yellow to black. His ten minutes were nearly gone. He touched the vial with white peanut milk in it. He hoped they were anxious to hear about that. His peanut milk had been written up in magazines like *Popular Mechanics.* "I see my ten minutes of time is about up. Of course I had to lose time in getting these samples out. . . ."

"We will give you more time!" cried Chairman Fordney in anticipation.

George moved his hand away from the vial of peanut milk and instead explained a compressed cake of hay and meal from

243

peanuts, used for livestock feed. Suddenly Illinois Representative Rainey, an opponent of tariffs, wanted to know why, if peanuts were so desirable, did they have to be protected? Wouldn't the demand raise their value? Several committeemen jumped to George's defense. A spirited discussion on tariffs and how they were used broke out among the members of the Committee. So much was at stake. Had George's presentation been derailed?

"Oh yes, sir," George finally grumbled, "that is all the tariff means — to put the other fellow out of business."

Everyone laughed.

Committee Chairman Fordney seemed elated with George. "Go ahead, brother. Your time is unlimited!"

George rewarded him. He held up the vial of white peanut milk. "Here is a bottle of milk that is extracted from peanuts. Now it is absolutely impossible to tell that milk from cow's milk in looks and general appearance." He held up a second bottle of white liquid. "This is normal milk."

He went on to explain how cream rose to the top on the peanut milk just as it did on cow's milk. He told them ice cream could be made from peanut milk. He showed them buttermilk made from peanuts. A quarrel broke out about the effect of peanut milk on the dairy industry. George was pleased, though. They wouldn't be discussing such a possibility if they hadn't taken his assertions very seriously. He went on to show them Worcestershire sauce, curds, cheese, various flavored punches, cooking oil, salad oil, face cream, ink, relish, mock chicken, and coffee — all made from peanuts. The Congressmen chuckled as George sprinkled in jokes about his experiences with farmers.

In all, he had talked nearly two hours!

He finished his presentation by quoting God's word to man in Genesis 1:29:

Behold, I have given you every herb bearing seed, which is upon the face of all the earth, and every tree, in the which is the fruit of a tree yielding seed; to you it shall be for meat.

Committee members then quizzed him on his education. Just in case none of them knew Iowa State was a premier school for agriculture he told them James Wilson, Secretary of Agriculture for sixteen years, had been his instructor for six years. George made sure the members knew he had shown only about half of the uses for the peanut. There were many more proven uses. And more uses were being developed all the time at Tuskegee Institute.

"You have rendered the Committee a great service," enthused Representative Carew.

John Garner of Texas exploded, "I think he is entitled to the thanks of the Committee!"[2]

Applause rippled through the Committee, then erupted into the gallery. George thanked God his presentation had gone so well. Later, he was glad to realize about the only thing he forgot to tell the Committee was that Henry Wallace, the incoming Secretary of Agriculture, had also been his instructor at Iowa State!

Weeks later a man from the United Peanut Association of America called George. "You completely won them over, Professor Carver. Before your talk the tariff was three-eighths of one cent for unshelled peanuts and three-fourths of one cent for shelled peanuts. The Ways and Means Committee recommended those tariffs be revised to three cents and four cents! That's an eight-fold increase in the tariff on unshelled peanuts!"

In his laboratory George worked not just on peanuts but on sweet potatoes and velvet beans too, as well as clays. In 1923 he received the Springarn Medal, an award intended to recognize achievement by a colored person. More and more George was asked to speak. His audiences were varied, from college groups to religious groups, from colored to white. George now spoke often at white colleges throughout the south. Even Booker Washington had never broken through that barrier. These southern colleges were closed to colored students. This exposed George to some bigots, but he bore the racial slurs. It was worth it if his talks made some white people realize that colored people could tackle any discipline a white person could tackle.

In late spring of 1924 he wrote a friend:

God has indeed been good to me and is yet opening up wonders and allowing me to peep in as it were. I do love the things God has created, both animate and inanimate. . . . I am not so good. I am just trying, through Christ, to be a better man each day . . .[2]

Sometimes opposition came from unexpected quarters. In a trip to New York in 1924 George spoke at the Marble Colle-

247

giate Church to the Women's Board of Domestic Missions of the Reformed Church of America. Accordingly, George spoke from the heart. His science came to him often in his laboratory as it had when as a child he saw a fancy lacework and then somehow duplicated it.

"God is going to reveal things to us that He never revealed before — if we put our hand in His." For emphasis George added, "No books ever go into my laboratory. The thing that I am to do and the way of doing it come to me. I never have to grope for methods. The method is revealed at the moment I am inspired to create something new. Without God to draw aside the curtain I would be helpless."[3]

If his audience was inspired, the *New York Times* was not. Two days later its editorial page blasted George. Divine inspiration? What an outrage! The editorial was headlined "Men of Science Never Talk That Way." Its judgment was merciless:

> . . . *It is therefore to be regretted that he should use language that reveals a complete lack of the scientific spirit. Real chemists do not scorn books and they do not ascribe their success, when they have any, to "inspiration." Talk of that sort will*

simply bring ridicule on an admirable institution and on the race. All who hear it will be inclined to doubt, perhaps unjustly, that Dr. Carver's chemistry [is] appreciably different from the astronomy of the once-famous Reverend John Jasper who so firmly believed that the sun went around a flat earth . . .[4]

George was appalled by the editor's lack of understanding. Of course he studied books. He had meant that once the books are mastered by the scientist the next step beyond the books requires inspiration. And his inspiration came from God. The remark that George brought ridicule upon Tuskegee and his own race hurt like a whiplash. It got worse. The *New York Times'* reprimand was picked up by even more unprincipled newspapers. Their headlines were cartoonish and simpleminded: "Negro Professor Aided by Heaven," "God His Book of Knowledge," "Colored Savant Credits Heaven," and "God Reveals Secret." What choice did George have but to ask the *New York Times* to print his explanation, or better yet to print a retraction for their simpleminded liberalism? He composed a lengthy letter:

249

I regret exceedingly that such a gross misunderstanding should arise as to what was meant by "Divine inspiration." Inspiration is never at variation with information; in fact, the more information one has, the greater will be the inspiration. Paul, the great scholar, says . . . in Galatians 1:12, "For I neither received it of man, neither was I taught it, but by the revelation of Jesus Christ." Many, many other equally strong passages could be cited . . .

George then went on in his letter to point out he had two degrees from Iowa State in Scientific Agriculture. His readings in chemistry were extensive. George listed some thirty authors who most influenced him in chemistry. He emphasized their books and papers were in his library. He did the same for the studies of botany and agriculture. He noted that he received many scientific journals. But he was not about to withdraw his belief in divine inspiration:

I thoroughly understand there are scientists to whom the world is merely the result of chemical forces or material electrons. I do not belong to this class. . . .

The master analyst needs no book; he is at liberty to take apart and put together substances,compatible or non-compatible to suit his own particular taste or fancy.

He furnished an example that surely even the liberals at the *New York Times* could understand:

I was struck with the large number of Taros and Yautias displayed in many of your markets . . . edible roots imported to this country. . . . Just as soon as I saw these luscious roots I marveled at the wonderful possibilities for their expansion. Dozens of things came to me while standing there looking at them. I would follow the same or similar lines I have pursued in developing products from the white potato. I know of no one who has ever worked with these roots in this way. I know of no book from which I can get this information, yet I will have no trouble in doing it. If this is not inspiration and information from a source greater than myself, or greater than any one has wrought up to the present time, kindly tell me what it is . . .[5]

The *New York Times* refused to print his

251

letter. Finally, friends took copies of his letter to other newspapers which did print it. George had endured yet one more form of discrimination. This time for his belief in a caring God.

Because of the storm of publicity over the Scopes Monkey Trial in Tennessee, where a teacher was prosecuted for teaching evolution, George was compelled to address the issue of evolution. The theory of evolution, although much more poorly substantiated than lay people realized, did not disturb George that much. He had found many fossils. Yet George believed man was created in the image of God, just as Genesis 1:27 stated. If in the first "days" of creation God wanted intermediary forms, "missing links," what did it matter? The real threat was the intolerance he had already suffered himself: that scientists became so materialistic they vehemently denied all other possibilities. "I am fearful lest our finite researches will be wholly unable to grasp the infinite details of creation, and therefore we lose the great truth of the creation of man,"[6] he wrote an inquirer.

In spite of the attack by the *New York Times* George enjoyed prestige now that he was past sixty. By 1927 he was touting even larger numbers of products for the vegetables he was investigating. The velvet bean yielded 35 products, the sweet potato 118 products, and the peanut — champion of them all — 200! The numbers of products from his laboratory increased daily. George also put out charts to show how superior peanuts were in protein content, what he called muscle builders, to everyday foods like bread, cheese, eggs, and beef. And to manual laborers calories were important. The greater the number of calories the more fuel for work!

"One ounce of peanuts has 160 calories," emphasized George. "The same amounts of beef or eggs have but 50."

Dr. Moton promoted George relentlessly. In 1930 sculptress Isabel Schulz was commissioned to do a bronze wall plaque for Tuskegee. George posed restlessly, even disagreeably, for his likeness. For its dedication at the 1931 commencement he was even more disagreeable. He refused to wear a cap and gown for the ceremony. Dr. Moton actually had to wrestle him into his garb. That, too, was like George — rare though it was — to suddenly do something completely juvenile. It reminded him of his time at Iowa State when he playfully chased the underclassman with a knife.

Often now people would walk right in his laboratory and gawk at him. Some white visitors would make coarse remarks about his race, thinking they were complimenting him. But he had many white friends around Tuskegee, too. Some supported him so openly he worried for their safety. But sometimes George could not see what was in the heart of a white person — until an incident erupted. At the burg of Notasulga, a farmer named Blanks had gone right to the back of the bus to sit with George.

Sure enough the white bus driver turned to face Blanks. "Y'all can't sit back there!"

Blanks' face reddened.

George was sweating plenty by then. . . .

fifteen

Blanks bellowed at the bus driver, "I'd sure enough rather walk than be driven by a bigot! Come on, let's get off the bus, Mr. Carver."

Sometimes love was hidden in the heart.

George now received over one hundred letters a day. He learned to read the last paragraph of a long letter to determine what the writer wanted, if anything. Otherwise he would never have been able to answer his mail. Sometimes he would gently reject a wild idea like turning dog droppings into perfume. Sometimes the writer would provoke tremendous nostalgia. "Are you the same George Carver that used to live here in Beeler?" a letter asked. And George would suddenly be overwhelmed by wonderful western Kansas memories. Some young folks — both white and colored — became his pen pals. He wrote, "Thank God I love humanity; complexion doesn't interest me one bit."[1]

One young white man sent George his

255

sketches to critique. George, who needed thirty-six hours in every day, still could not neglect a fellow artist. "The shadows in sketch number one extend too far around for your ball to float out in the air," he muttered as he scribbled his answer. "Hmmm, sketch number two has wonderful strength and color. Oh, no. What's this? Number three is all foreground, absolutely no middle distance or extreme distance, which are so essential to a pleasing picture."

His critique was bluntly honest, but he lavished praise on the young man for his efforts. He never hesitated to impart to his correspondents what he called "the true secret of success, happiness, and power," embodied for him in four Bible passages:

Not that we are sufficient of ourselves to think any thing as of ourselves; but our sufficiency is of God; Who also hath made us able ministers of the new testament; not of the letter, but of the spirit: for the letter killeth, but the spirit giveth life. (2 Corinthians 3:5–6)

Study to show thyself approved unto God, a workman that needeth not to be ashamed, rightly dividing the word of truth. (2 Timothy 2:15)

In all thy ways acknowledge him, and he shall direct thy paths. (Proverbs 3:6)

I can do all things through Christ which strengtheneth me. (Philippians 4:13)

Yes, George had tested these profound words, even doubted a time or two. But his belief had become rock solid.

His letters were full of love and fatherly advice:

. . . have a good time socially. You need it. And by all means don't let worry weight you down. I have seen so many students (old young men) who had let the "tons" of worry of which you so aptly speak weight them down and make them look old, act old and feel old . . .[2]

For years George had written his Iowa State friends. By 1931 James Wilson, the older Henry Wallace, and Dr. Pammel had passed away. George remembered Dr. Pammel with great fondness and wrote his daughter Violet, "O, how I love him, marvelous man that he was."[3] Others were still

257

alive and very prominent. In 1932 young Henry Wallace, the boy who had toddled after George on his collecting trips, became Secretary of Agriculture under President Franklin Roosevelt.

Tuskegee Institute now provided George, nearly seventy, an aide, Harry Abbott. Abbott made all arrangements, carried luggage, and provided real companionship. George had not been so attached to anyone since his brother Jim. They even developed a brotherly relationship, best of confidants but quarrelsome, too. President Moton's health suddenly declined and he had to retire in 1935. The attitude of his successor and his lieutenants toward George was pragmatic. They regarded his reputation as a valuable resource to Tuskegee. But George felt they did not appreciate his scientific prowess. When George's confidant Harry Abbott took a job in Chicago, the new president of the Institute replaced him not with an aide but a laboratory assistant. George was cool to chemist Austin Curtis when he first arrived. It certainly wasn't his idea. But the even-tempered competence of Curtis won him over. Soon George joshed with Curtis as freely as he ever had in his best relationships with friends.

"Sir," said Curtis, "they say I grow more like you every day."

"Oh phooey, you don't resemble me at all."

"They call me 'Baby Carver,' sir."

"If you keep on aggravating me, Baby Carver, I'll have to give you a good whipping. Your backside won't hold shucks when I get through. You'll be crying some big elephant tears then."

George not only traveled less but turned more of the laboratory work over to Austin Curtis. George remained a national figure, though. There was scarcely a prominent magazine that hadn't written a story on him. Thomas Edison offered him a job that he declined. Automobile tycoon Henry Ford corresponded with him and invited him to Michigan and to his estate in Georgia. President Franklin Roosevelt stopped by with his entourage on the way to his getaway at Warm Springs, Georgia. Two different movie studios sent representatives to the Institute to talk to George about a movie of his life. One was from Metro Goldwyn Mayer.

"He was absolutely overcome with emotion," said George later in amazement.

Beginning in 1938 George's health became so bad he rarely traveled. At

259

seventy-five he focused on preparing a museum at the Institute. He was anxious to preserve his many years of work. A fine new building was designed, then faltered because of no funds. Finally the trustees offered him the one-story red-brick laundry building, which would be modified for George's needs. In failing health George felt he had no time to quibble. So he accepted the offered building, barely adequate though it was, and he himself was moved into nearby Dorothy Hall, a guest house. Henry Ford had an elevator installed for George at the guest house so he would not have to struggle up and down the stairs. Now he hobbled back and forth between the guest house and the museum, devoting his time to preserving his specimens and papers and canvases. Beyond preservation he was anxious to arrange his treasures, so they were not just discombobulated remnants of a long career but groupings that would teach visitors. He had accumulated so much: two-foot-high jars of vegetables, soil samples, ceramics, stained boards, dyes, pigments, over one hundred sweet potato products, three hundred peanut products, fabrics, rugs, and wallpaper designs. Oh Lord, he even had the skeleton of Betsy, the very first ox to plow the experiment station plots! On July 25, 1939, the museum officially opened, but more work remained. His canvases and fancy works had yet to be displayed in a separate gallery.

"They must be done just so," George explained to Curtis.

When he occasionally traveled he was accompanied by Curtis. "Baby Carver" was not as compliant as George was. In September 1939 they were refused a room in a New York hotel, even though they had reservations. Curtis simply sat a very exhausted George down in a comfortable chair and negotiated. He was finally forced to call the *New York Times*, *New York Post*, *Pittsburgh Courier*, and *Chicago Defender*. With great reluctance the hotel gave them one of many rooms that were available. The hotel lost the battle and the war, because newspapers all over the country ran stories of their racial prejudice. George would never had caused such a ruckus. Curtis also monitored press coverage of George and complained sharply against stereotypical slurs like "shuffling" and "old darky." George was not totally compliant, though; he had once told a white entrepreneur that it was not only a poor business decision to use the brand name Pickaninny

261

Peanuts, but that colored people were very offended by such a caricature.

But most intriguing to George at this time of his life in the late 1930s were all the biographies in progress. . . .

sixteen

George was deluged with writers gathering material for their various biographies of him. In his more honest moments he knew the rush was on because he was now well past seventy and often seriously ill. Still, it would have been more depressing if no one had been interested at all. Lucy Crisp interviewed him as did Jessie Guzman, but Mrs. Rackham Holt impressed him most. She had a book contract with Doubleday already signed. He read one of her early drafts:

By the time he was virtually a baby in the woods he wanted to know the name of every stone, insect, flower he saw. He had a book given him by Aunt Sue — Webster's old blue-back Speller — that had a picture of a man climbing a high cliff on the top of which stood a temple of learning. Few people thereabouts could even write, but George had studied the speller until he knew every word. However, it did not reveal the names of the birds, so he made up names to suit him-self. Having once tasted the fruit of knowledge and caught a glimpse of the mysteries hidden in words, he could not rest content.[1]

"Beautiful," commented George. "I can't put it down."

To George she wrote as well as Carl Sandburg. He had once met the renowned Sandburg in New York. Imagine, George was going to have a biographer like the one who wrote about Abraham Lincoln. Although her depiction of George was not only skilled but very flattering, she moved at a much slower pace than George would have. And she wanted to know too much. "There are some things that an orphan child does not want to remember,"[2] he complained. Must she dredge up his memory of the Fort Scott lynching? The rejection at Highland College?

The world forged ahead. Some news was good. By 1940 peanuts were the second largest cash crop in the entire south. George had some hand in that. Franklin Roosevelt was reelected, with Henry Wallace as his vice president. Imagine, George's toddling little nature pal was vice president! But the rest of the news was like

George's health: rotten and getting worse every day. A terrible fascist dictator was conquering Europe. Japan had attacked America's naval base in Hawaii. Men all over America were being enlisted to fight. George prayed that it would not be another great war, but if it was, God, please let the colored soldiers be treated right this time. Perhaps with Franklin Roosevelt in office they would be.

"The Lord knows I don't have much fight left."

By 1941 he had opened the gallery portion of the museum, well satisfied with the arrangement of his paintings and fancy works. Some people again rudely gawked at him. And yes, often they would remark coarsely how much more amazing it all was that he was colored. George liked people though, and many encounters were pleasant. B. B. Walcott, who would write of her experience in a 1942 issue of *Service*, asked him a question he had heard ten thousand times: "How can you do so many different things?"

"Would it surprise you if I told you I haven't been doing many *different* things?" he replied patiently. Then he quoted Lord Tennyson's poem "Flower in the Crannied Wall":

265

Flower in the crannied wall
  I pluck you out of the crannies
I hold you here, root and all, in my hand,
  Little flower — but if I could
    understand
What you are, root and all, and all in all,
  I should know what God and man is.[3]

"Writing, painting, crocheting, music, experimenting — these are all the same thing: a search for truth," he said. Sometimes he made his point with John 8:32, one of his favorite Bible passages, " 'And ye shall know the truth, and the truth shall make you free.' Art and science both spring from a search for truth."

Meanwhile, the book by Rackham Holt had him by the throat. Eighteen schools had been named after him and yet he let this artistic statement of the meaning of his life become his chief aggravation. "What is the delay?" he fumed. His time on this earth was limited. He fretted over his old age and his waning days. Often he scolded Harry Abbott in his letters if Abbott had not answered a letter quickly enough. Not being able to write any longer with a crippled right hand and needing a stenographer didn't help his outlook, either. But what was delaying Mrs. Holt's book? When she requested additional material on him after working on the book three years, George's patience crumbled. On October 14, 1942, he wrote her in anguish:

I wish so much that the book could be finished. I think that it should be closed up. . . . I was hoping so much that this book could be finished before it had to close with something sordid . . .[4]

There! He said it as directly as he could. Mrs. Holt, please finish the book before I die! But by New Year's 1943 the book was still not published. George had fallen in December while opening a door and the inactivity that resulted from the enforced bed rest gravely weakened him. He had bounced back a dozen times before. But this time was different. His whole body was as sore as a boil. He could barely lift his little finger. This time he wasn't going to bounce back, he realized. Well, why should he? He could do no more laboratory work. He hadn't done any significant research since 1938. He couldn't write letters. The museum was at last arranged the way he wanted it. It was time for his reward. He had been impatient with Mrs. Holt, but she had forced him to remember

267

all the terrible things the Lord had pulled him through. Yes, George was one lucky fellow the way the Lord directed his path. With God's help George had changed a few things, too.

Years before he had written a pen pal:

There are times when I am surely tried and compelled to hide away with Jesus for strength to overcome. God alone knows what I have suffered, in trying to do as best I could the job He has given me in trust to do. Most of the time I had to work without the sympathy or support of those with whom I associated. Many are the strange paths God led me into. . . . God has so willed it that there were always a few good friends to encourage me and strengthen me when the burden seemed greater than I could bear . . .[5]

On January 5, 1943, George Washington Carver joined the Lord.

endnotes

Chapter 3
1. Neider, Charles (ed.). *The Complete Short Stories of Mark Twain*. New York: Bantam Books, 1957, p. 67.

Chapter 9
1. Millbrook, Minnie Dubbs. *History of Ness, Western County, Kansas*. Detroit: 1955, p. 265.
2. Holt, Rackham. *George Washington Carver: An American Biography*. New York: Doubleday and Co., 1943, p. 53.
3. Holt, p. 54.

Chapter 10
1. Holt, p. 67.
2. Holt, p. 68.
3. Holt, p. 69.
4. Reprinted from *George Washington Carver: In His Own Words* edited by Gary R. Kremer, by permission of the University of Missouri Press. Copyright© 1987 by the Curators of

the University of Missouri.
5. Kremer, p. 45.
6. Holt, p. 71.

Chapter 11
1. McMurry, Linda O. *George Washington Carver: Scientist and Symbol.* New York: Oxford University Press, 1981, p. 38.
2. McMurry, p. 39.
3. Holt, p. 96.
4. Holt, p. 97.
5. McMurry, p. 44.
6. McMurry, p. 150.

Chapter 12
1. Kremer, p. 47.
2. McMurry, pp. 60–61.
3. McMurry, p. 71.
4. Kremer, p. 138.
5. Kremer, p. 76.

Chapter 13
1. Holt, pp. 226–227.
2. Holt, p. 235.
3. McMurry, p. 110.
4. Holt, p. 254.

Chapter 14
1. Kremer, pp. 103–113.

2. Kremer, p. 37.
3. Holt, p. 265.
4. Holt, p. 266.
5. Kremer, pp. 128–130.
6. Kremer, p. 133.

Chapter 15
1. Kremer, pp. 131–132.
2. Kremer, pp. 183–184.
3. Kremer, p. 58.

Chapter 16
1. Holt, p. 21.
2. Kremer, p. 31.
3. Williams, Oscar (editor). *Immortal Poems of the English Language*. New York: Pocket Books, 1952, p. 375.
4. Kremer, pp. 32–33.
5. Kremer, p. 172.

bibliography

I. Two biographies are particularly noteworthy:
   Holt, Rackman. *George Washington Carver: An American Biography.* New York: Doubleday and Co., 1943.
   McMurry, Linda O. *George Washington Carver: Scientist and Symbol.* New York: Oxford University Press, 1981.

II. Correspondence of George Washington Carver:
   Kremer, G. R. (ed.). *George Washington Carver: In His Own Words.* Columbia: University of Missouri Press, 1987.

III. Other relevant reading:
   Calhoun, William G. (ed.). *Fort Scott: A Pictorial History.* Bourbon County Historic Preservation Assoc. 1978.
   Centennial Committee. *History of*

Ottawa County, Kansas, 1864–1984. Centennial Committee, 1984.

Dick, Everett. *The Sod-house Frontier, 1854–1890.* New York: D. Appleton-Century Company, 1937.

Millbrook, Minnie Dubbs. *History of Ness, Western County, Kansas.* Detroit: 1955.

Miner, Craig. *West of Wichita: Settling the High Plains of Kansas, 1865–1890.* University Press of Kansas, 1986.

Washington, Booker T. *My Larger Education.* New York: Doubleday & Company, Inc., 1911.

Washington, Booker T. *Up From Slavery.* New York: Doubleday & Company, Inc., 1900.

Yearbook of Agriculture, After A Hundred Years. Washington, D.C.: U. S. Dept. Agriculture, 1962.

The employees of Thorndike Press hope you have enjoyed this Large Print book. All our Large Print titles are designed for easy reading, and all our books are made to last. Other Thorndike Press Large Print books are available at your library, through selected bookstores, or directly from the publisher.

For more information about titles, please call:

(800) 223-1244
(800) 223-6121

To share your comments, please write:

Publisher
Thorndike Press
P.O. Box 159
Thorndike, Maine 04986